Mastering the Construction Startup

Mastering the Construction Startup

A Business Infrastructure Guide

Nick B. Ganaway

Copyright © 2025 by John Wiley & Sons Inc. All rights reserved, including rights for text and data mining and training of artificial intelligence technologies or similar technologies.

Published by John Wiley & Sons, Inc., Hoboken, New Jersey.

Published simultaneously in Canada.

No part of this publication may be reproduced, stored in a retrieval system, or transmitted in any form or by any means, electronic, mechanical, photocopying, recording, scanning, or otherwise, except as permitted under Section 107 or 108 of the 1976 United States Copyright Act, without either the prior written permission of the Publisher, or authorization through payment of the appropriate per-copy fee to the Copyright Clearance Center, Inc., 222 Rosewood Drive, Danvers, MA 01923, (978) 750-8400, fax (978) 750-4470, or on the web at www.copyright.com. Requests to the Publisher for permission should be addressed to the Permissions Department, John Wiley & Sons, Inc., 111 River Street, Hoboken, NJ 07030, (201) 748-6011, fax (201) 748-6008, or online at http://www.wiley.com/go/permission.

The manufacturer's authorized representative according to the EU General Product Safety Regulation is Wiley-VCH GmbH, Boschstr. 12, 69469 Weinheim, Germany, e-mail: Product_Safety@wiley.com.

Trademarks: Wiley and the Wiley logo are trademarks or registered trademarks of John Wiley & Sons, Inc. and/or its affiliates in the United States and other countries and may not be used without written permission. All other trademarks are the property of their respective owners. John Wiley & Sons, Inc. is not associated with any product or vendor mentioned in this book.

Limit of Liability/Disclaimer of Warranty: While the publisher and author have used their best efforts in preparing this book, they make no representations or warranties with respect to the accuracy or completeness of the contents of this book and specifically disclaim any implied warranties of merchantability or fitness for a particular purpose. No warranty may be created or extended by sales representatives or written sales materials. The advice and strategies contained herein may not be suitable for your situation. You should consult with a professional where appropriate. Further, readers should be aware that websites listed in this work may have changed or disappeared between when this work was written and when it is read. Neither the publisher nor authors shall be liable for any loss of profit or any other commercial damages, including but not limited to special, incidental, consequential, or other damages.

For general information on our other products and services or for technical support, please contact our Customer Care Department within the United States at (800) 762-2974, outside the United States at (317) 572-3993 or fax (317) 572-4002.

Wiley also publishes its books in a variety of electronic formats. Some content that appears in print may not be available in electronic formats. For more information about Wiley products, visit our web site at www.wiley.com.

Library of Congress Control Number: 2025906748

Print ISBN: 9781394292325
ePDF ISBN: 9781394292349
ePub ISBN: 9781394292332
oBook ISBN: 9781394292356

Cover Design: Wiley
Cover Image: Generated with AI using ChatGPT 4o

Set in 12.5/14.5pt STIXTwoText by Lumina Datamatics

SKY10104394_042525

For my loyal employees past and present, who made it all possible.

Acknowledgments

I have bounced various business ideas off my friend Jim Nellis for the past 35 years, usually over a glass of wine, and I always come away wiser than before. Jim's comments and suggestions during his reading of a couple of early drafts of this book were right on target and a vital contribution to its value. Thanks, Jim, as always.

It is only in the most positive terms that I can express my highest regard and appreciation for Tim C. Taylor. Tim, whose home is in England, is an accomplished author and editor who has been my writing guru since 2012, patiently holding my hand through every aspect of writing, editing, and publishing. Thank you, Tim. Your teaching, guidance, and friendship have been and continue to be of irreplaceable value to me.

I am forever grateful to my many friends, business associates, and employees who have traveled this road with me. Their wisdom is reflected in these pages. You know who you are. Thank you.

Finally, I want to express my love and gratitude to my children, Ginger Ganaway Smith, Julie Ganaway Blanchard, and John Ganaway, for their ear, eternal encouragement, and love in this and all of my writing and business life. Special thanks to Julie for graphics design. Love you, kiddoes.

Author's Disclaimer

I am a businessman, not a lawyer, accountant, or insurance professional. The subject matter, advice, and recommendations provided in this book are based on my business experience and interaction with my various professionals over many years in relation to my businesses and are not intended to nor should they be construed as legal or other advice that may be obtained from licensed professionals.

About the Author

Nick B. Ganaway founded and operated Ganaway Contracting Company for 25 years, a commercial general contracting firm operating in the Southeastern US, before selling to its present owner. The infrastructure he built for his company has supported it through the best and worst of times. Ganaway Contracting Company celebrated its 50th anniversary in 2024.

Nick learned hard work from an early age on the small Texas farm where he grew up. After graduation from college, he worked for seven years in a major US company where he learned how business organizations work and their dependence on infrastructure that is built and continually nourished from their beginnings.

Nick's business writings and articles have appeared in print and online. He is also the author of *Construction Business Management: What Every Commercial Contractor, Builder & Subcontractor Needs to Know* (Wiley).

He graduated from the University of Texas at Arlington with a B.S. in Industrial Engineering. He lives in Atlanta and has three adult children and six grandchildren.

Contents

Preface   *xxi*
Introduction   *xxvii*

Part 1   The Contractor   *1*

1 Entrepreneurial Characteristics   *3*
It's Not Easy   *4*
Your Dual Role as Owner and Manager of Your Company   *6*
Passion for Your New Baby Is Good, but...   *7*

2 Company Culture   *9*

3 Elements of Leadership   *11*
Vision and Strategic Direction   *11*
Emotional Intelligence   *12*
Communication Skills   *12*
Adaptability and Resilience   *13*
Integrity and Ethical Leadership   *13*
Team Development and Empowerment   *13*

xiv | *Contents*

Part 2 Regulatory Matters *15*

4 Business and Government Regulations *17*

Incorporating Your Company *17*
Registration in Other States *17*
Registered Agent *18*
Professional Licensing *18*
Municipal Business Licensing and Permitting *18*
Corporations and LLCs *19*
The LLC *21*
The S Corporation *22*
The C Corporation *22*
Sole Proprietorship *23*
Partnership *24*
Partnership Agreement *24*
Establishing Your Corporate Entity *25*
Choosing the Name for Your Company *25*

Part 3 Getting It Done *27*

5 The General Contractor *29*

The Owner–Contractor Relationship *30*
Managing Risk *32*
Construction Type *34*
Union or Nonunion *34*
Managing the Common Causes of Contractor Failure *35*
Increase in Project Size *35*
Changing Your Geographic Area *36*
Change in Key Personnel *36*
Taking on a New Type of Construction *36*
Lack of Management Maturity *36*
Task Dependency Software *38*
Construction Document Software *38*
Construction Bookkeeping Software *38*
Awareness of the Economy and Other Business Threats and
Opportunities *39*

Contents | **xv**

Negotiation *39*
Proactively Manage Budgets *40*
Demand High Quality *41*
Cultivate Relationships *41*
Meet the Schedule *42*
Stakeholder Communication *42*
Human Resource Management *42*
Trust *43*
Marketing *43*

6 Subcontractor Management *47*
Effective Subcontract Documents *47*
   Trust *48*
   Prequalification of Subcontractors *49*
   Subcontractor Proposal and Scope of Work *50*
   Notice to Proceed Precaution *51*
   Official Notice *51*
   Inspection of the Subcontractor Work *52*
   Topographical Site Survey *52*
   Differing Conditions *53*
   Independent Contractor Versus Payroll Employee *55*
   IRS Regulations for Independent Contractors *56*
   Managing Change Orders *57*
   Effective Change Order Procedures *58*

7 Construction Disputes *61*
Design Defects *61*
Workmanship Defects *62*
Negligence *62*
Changes in Scope *63*
Changes in the Work *63*
Contract Terms and Conditions *63*
Payment *64*
Methods of Dispute Resolution *64*
   Negotiation *64*
   Mediation *64*
   Arbitration *65*

xvi | Contents

8 Project Delay *67*
Causes for Delay *67*
Be Prepared *69*

9 Contracts and Agreements *71*
What Is a Contract? *71*
The Value of a Written Agreement *72*
Types of Construction Contracts *73*
Lump-sum Contract *73*
Unit Price Contract *74*
Cost-plus Contract *74*
Incentive Contract *74*
Design–Build Contract *75*
Getting Paid *76*
Example Lawsuit *79*

Part 4   Resources *83*

10 Lawyers *85*
Construction Industry Lawyer *86*
General Practice or Business Lawyer *87*
Employment Lawyer *87*
Workers Compensation Lawyer *87*
Real Estate Lawyer *88*
Mergers & Acquisitions Lawyer *88*
Civil Litigation Lawyer *88*
Tax Lawyer *88*
Intellectual Property Lawyer *89*
Personal Injury Lawyer *89*
Immigration Lawyer *89*
Estate Planning Lawyer *89*
Deal Making *89*

11 Accounting and Taxes *91*
Tax Preparation *92*
The Income Statement (Profit and Loss Statement) *94*
The Balance Sheet *95*
Statement of Cash Flows *95*
Summary *96*

Contents | **xvii**

12 Hiring the Right People *97*
So, Who Are the Right People? *98*
Tips for Interviewing Prospective Employees *99*
Job Benefits *100*
Jobsite Facilities and Conveniences *102*
Onboarding New Employees *104*
Employee Meetings *105*
Employee Handbook *105*
Noncompete Agreement *106*

13 Banking *109*
Loan Documentation *110*
Personal Guarantee *111*
Variable Rate Loans *111*
Credit Unions *111*
Private Lenders *112*

14 Construction Insurance *113*
General/Commercial Liability Insurance *114*
Workers' Compensation Insurance *115*
Employer Responsibilities Under Workers Comp Insurance *116*
Workers' Comp Insurance Benefits to Employers *116*
Cost of Workers' Comp Insurance *116*
Builders' Risk Insurance *117*
Business Owner's Policy *117*
Product Liability Insurance *117*
Professional Liability Insurance *118*
Commercial Property Insurance *118*
Fidelity Insurance *118*
Home-based Business Insurance *118*
Coinsurance *119*
Making Claims *120*
Performance and Payment Bonds *120*
Performance Bonds *121*
Payment Bonds *121*
Summary *121*

xviii | *Contents*

15 Business Plan *123*
Purpose and Evaluation *123*
Presentation for the User *123*
Proposal Guidelines *124*

Part 5 Ideas *127*

16 Niche Contracting *129*
Advantages of Niche Contracting *129*
Additional Niche Advantages *131*
Chain Store Owners and Franchisors *133*

17 Outside Board of Advisors *137*
Is an Advisory Board Necessary? *138*

18 Case Study of an Existential Business Crisis – A Personal Account *141*
Another Shoe Drops *142*
A Perfect Storm *142*
Looking Back *143*
A Tough Decision *144*
The Personal Effect *144*
Success, at Last *145*

19 Investing in Real Estate as a Parallel Business *147*
Owning STNL Properties – a Successful Proven Strategy *148*
General Characteristics of STNL Properties *148*
The STNL Marketplace *150*
Primary Factors in Pricing an STNL Property *151*
Replacing the Property Early *153*
STNL Properties Versus the Stock Market *154*
The Case for Investing in Real Estate *155*
Dealing with Capital Gains Tax *155*
The 1031 Tax-deferred Exchange *156*
In Conclusion *156*

Contents | **xix**

20 Summary Checklist for Startup Businesses *159*
What to Do First *159*

21 Useful Reading *161*

Appendix Contents *165*

    A Partnership Agreement Example *167*
    B Non-compete Agreement Example *171*
    C Business Loan Proposal Form Example *173*
    D Business Plan Example 1 *175*
    E Board of Advisors Agreement Example *181*
    F Employee Handbook Example *185*
    G Independent Contractor Agreement Form Example *191*
    H Business Plan Example 2 *197*
    I ACORD Certificate of Insurance Form Example *201*

Index *203*

Preface

You're Not Alone

I have written *Mastering the Construction Startup: A Business Infrastructure Guide,* for the man and woman optimistically starting and owning their own business but confronting the nagging uncertainties I faced decades ago – exciting, but also a little scary. It's exciting because it is the fulfilment of our dream of owning our own business and running it the way we have imagined, of being our own boss while knowing it is not without considerable risk. Because we believe we can make our and our families' lives become more fulfilling. And because we have a twitch that can be calmed only by making life happen *for* ourselves, rather than allowing it *to* happen *to* us.

Taking advantage of the principles and experiences described in this book puts you in a much better place than I was when I began. At that time I had been responsible for hiring and supervising contractors for Shell Oil's light commercial construction projects for several years, but upon becoming a contractor I quickly learned there is a wide chasm between the requirements for a project owner's representative and those for a general contractor in terms of knowledge, skills, resources, risk, and hard-earned experience.

There was no Amazon or Barnes & Noble nor did I find a practical book to help me prepare for the new world I was entering. I found myself caught between the fear of uncertainty and the determination for success when I was awarded the first job I bid on as a general contractor, which was a package contract to build two restaurant projects to start immediately and complete both of them simultaneously. I was prepared at best to win just one of the projects.

I had negligible personal net worth and a little US Small Business Administration money for working capital. The two jobs were separated by a 40-minute drive across the center of the Atlanta metro area. I was the project xxii | *Preface*

manager and superintendent for both projects as well as billpayer and materials procurer. To meet the start schedule, I had to establish credit with suppliers on the fly. I had no lawyer when I signed the contract the owner put in front of me.

Of course, I had no subcontractor relationships and the subcontractors I contacted were naturally skeptical of this newbie. In busy times, subcontractors stick with contractors from whom they're likely to get continuing work.

At this point, I had committed just about every contractor sin that usually leads to failure. But the angels were with me.

Many of the principles described in this book are derived from my experience gained from starting and running my general contracting firm for 25 years, and they offer contractors in their startup and early years the information and tools developed and honed by my experience since those early days.

What You Can Learn

Below are just some of the empowering bits of knowledge you can learn from this book.

- The elements of construction business infrastructure and why they must be in place on your first day in business
- Essential characteristics of a general contractor
- Duties and responsibilities of the general contractor
- The failure-prone choices contractors often make
- Your essential role in the development of your company's culture
- Elements of leadership
- Elements of risk and how to manage them
- Union vs nonunion projects
- Metrics that lead to successful construction projects
- Subcontractor management
- Honing your communication skills
- The value of integrity and ethical leadership
- Managing differing conditions (contract conditions vs actual conditions)
- The common causes for contractor failure
- Reasons for extensive documentation
- Controlling the risk posed by change orders
- Managing construction disputes
- Independent contractors vs payroll employees

Preface | **xxiii**

- Methods of conflict resolution
- The role of negligence in construction defects
- The essential role of your business plan and how to prepare it
- Types of business legal structure for your company
- Registering your company to do business
- Professional licensing
- The inherent risk of sole proprietorship
- How to structure business partnerships
- The elements of a partnership agreement
- Choosing the name for your company
- Hiring the right people for your company
- Onboarding new employees
- Construction jobsite facilities
- The employee handbook
- The controversial noncompete agreement
- Record keeping
- The basic components of a set of financial statements
- Choosing your legal, accounting, and insurance professionals
- The different legal specialties
- The personal guarantee and what to do when it is required
- The components of a legal contract
- Types of construction contracts and agreements
- Steps to ensure that you get paid for your work
- Types of business insurance
- The complexity of builder's risk insurance
- The risk of becoming co-insurer with your insurance carrier
- Performance and payment bonds
- Your business insurance carrier – whose side is it on, anyway?
- Establishing banking relationships
- Variable rate vs fixed rate loans
- Documenting your loan application
- The advantages of niche contracting
- Considering a board of advisors
- A case study: the author's three-year business nightmare and recovery
- Checklist for startup businesses
- Essential form and checklists

Why Read This Book?

Most books written for construction business startups and entrepreneurs are written by people who sit in paneled offices, who attend construction business seminars, who read construction magazines, but who never started or ran a successful construction company of their own. Their books discuss various aspects of contractors and construction, but they rarely mention *business infrastructure*, which is the essential first order of business.

This is not one of those books.

And this is not one of those authors.

What Is Business Infrastructure?

Your businesses' infrastructure interconnects your organization's goals with the essential elements of people, systems, processes, culture, and tools required for profitable and sustainable growth. These five elements are the underpinnings upon which all of your business activities and strategies will be built.

In this book, you will learn these elements of the supportive infrastructure critical to every business organization and have them in place when you open the doors on Day 1. Like the security system in your home, your business infrastructure sits quietly in the background while you go about your business, knowing the resources you need to effectively manage your business are at your fingertips.

Left unmanaged, your infrastructure will evolve on its own but that would be like, say, letting a stranger hire your key employees for you.

If you want the elements of your organization's infrastructure to reflect your style, your likes and dislikes, and your company's culture, they must be identified and vetted by you and only you.

Every entrepreneur who has experienced the losses and the wins, the caveats, the mistakes, the heartaches, the sweat, the tears, the thrills... of starting and successfully guiding their own construction business to success has gotten there on the shoulders of the solid foundation underlying their organization: *its infrastructure.*

When I was preparing to start my construction company decades ago, I had never thought about something called business infrastructure. This lack of knowledge led to misdirection and disorganization during this riskiest time in the life of any new business. My survival marked the beginning of my decades-long study of business infrastructure.

You have dreamed of stepping into construction for yourself for a long time. Studied the competition, decided on the type of work you want to pursue, developed your growth strategies, lined up your financing, negotiated with your vendors, and fine-tuned the services you will offer. Now protect your investment by building a strong infrastructure for your organization.

Introduction

All businesses start small. Think of Ford, Dell, Amazon, and Joe's Bicycle Repair Shop. These companies, like most general contracting and subcontracting firms, were started by men and women such as you and me who had an idea and pursued it with determination, struggle, and hardship. Strong leaders build strength and durability into their company's infrastructure, and not one of these businesses would have survived without that in place.

Three of these examples grew into industry giants, yet most new businesses remain small like Joe's.

Consider these statistics. According to the US Small Business Administration Office of Advocacy, employer firms with fewer than 20 employees made up 89% of all US businesses, and those with fewer than 10 employees accounted for 79%. Perhaps surprisingly, small businesses accounted for about 44% of US GDP, the measure of all goods and services produced over a given period of time. Nearly half (47%) of all US employees work for a small business, yet 81% of small businesses have no employees, meaning the owner does it all.

More so than startups in many fields, choosing to become an entrepreneur in construction is a gutsy decision and career path that takes confidence, commitment, hard work and information. According to the US Bureau of Labor Statistics, in the 25-year period through 2019 an average of 67.6% of new business startups survived at least two years. The five-year survival rate was 48.9%.

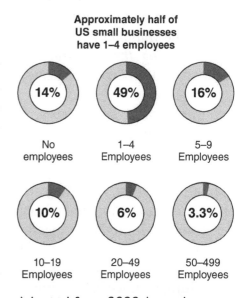

Adapted from 2022 Annual Business Survey, Census Bureau.

xxviii | *Introduction*

But don't stop reading here.

Typical causes for this dismal failure rate, anecdotally reported by contractors and other business owners, include:

- Insufficient working capital
- Poor management decisions
- Wrong or poorly researched target market
- Bad partnerships
- Ineffective marketing
- Any combination of these

As we were told in some math class, a problem well defined is half solved. *Mastering the Construction Startup: A Business Infrastructure Guide* offers you the keys to convert this essential knowledge into practical application within your company, which, in turn, will increase the likelihood of your business success exponentially.

When you start your own business, you become part of an essential, highly respected, and rewarding part of the US economy. According to a 2017 study by Fidelity Investments, as reported in US News and World Report, 88% of millionaires are self-made. This of course does not mean that all self-made entrepreneurs are millionaires. On the contrary, the army of successful and fulfilled businessmen and women who are *not yet* millionaires by far outnumber those who are. And, the journey to achieving any goal can be as rewarding and fulfilling as the destination, if not more so. (The author Thomas J. Stanley explores this topic in depth in his book, *The Millionaire Next Door*, and his other well-researched books.)

Mastering the Construction Startup: A Business Infrastructure Guide will prepare you to become a successful member of the group of men and women who have chosen to be their own boss after spending time or maybe their first career in businesses or organizations with layers of management, endless meetings, shared cubicles, internal politics, union rules, and/or difficult bosses.

Or maybe you've just graduated from college or finished other training and know you want to do something on your own. A version of that scenario explains my first venture into owning my own business. Right out of college, I landed a rewarding job with a major corporation in which I had a high degree of autonomy and the expectation for continuing advancement. But after my seventh year, I knew corporate life was not for me. It was an invaluable experience, but I wanted to do it for myself, which is likely the reason you are taking the plunge. I have been employed solely within my own businesses for the past several decades.

Part 1

The Contractor

Part 2

The Contractor

1

Entrepreneurial Characteristics

Entrepreneurs are people who take a *unique* or *innovative* idea and run with it. Some of them are serial entrepreneurs who often move on to their next challenge.

By contrast, other businessmen and women open their own new business not to innovate but to make their way in an established field such as construction.

In this book, I use the term "entrepreneur" interchangeably with "businessperson" to describe all businessmen and women who started and ran their own businesses, including those who started as an innovator.

A study published in the *Journal of Business Venturing* in March 2019 compared the brain patterns of 21 entrepreneurs and 21 parents who were not entrepreneurs. The study was designed to investigate how and why company founders bond with their business creation.

As reported in the *Wall Street Journal*, this study using functional magnetic resonance imaging (fMRI) found that when entrepreneurs think about their businesses, their brain patterns are very similar to the brain patterns of parents when they think about their children.

As parent, business founder, and operator for several decades, I am not surprised by these results. There really is nothing in your life other than your loved ones that is as dear and meaningful to you as the company you dreamed about, risked your life savings for, and worked impossible hours to finally bring to fruition and beyond.

Mastering the Construction Startup: A Business Infrastructure Guide, First Edition. Nick B. Ganaway.
© 2025 John Wiley & Sons, Inc. Published 2025 by John Wiley & Sons, Inc.

4 | *1 Entrepreneurial Characteristics*

It's Not Easy

You are probably used to working hard and putting in a lot of hours at your present or former job, but it is likely that starting or acquiring and running your own construction business is going to require even more of you. Not only in terms of time and effort, but also in many unfamiliar challenges as you travel new avenues, all the while knowing you have a lot or maybe everything on the line.

However, you will have a lot of company, including everyone in the past, present, and future who gets that impossible-to-ignore drive to take their future into their own hands and accept the same risks you are taking.

You might have thought you'll no longer have a boss, but you will in fact have the most demanding one you've ever had. He or she will be sitting right there on your shoulder asking hard questions. Judging your performance. Waking you up at 3 a.m. to question whether you put the right markup on the bid you submitted yesterday. Whether hiring Jack over Josh as your new project manager was the right decision. Demanding that you excel in every decision you make.

And you will at some point ask yourself, *"What was I possibly thinking?"* when you made this choice, even though you know the answer. But when those who never experienced the entrepreneurial fire in the belly or took total responsibility for their own fate notice the outward things it's costing you now – working long hours, skipping date night too often, etc. – they won't understand what drives you.

What you will get in return for your extraordinary sacrifices is the pride, the excitement, the freedom, the joy, and the challenge in watching your construction business grow and in knowing that *you* made it happen. And the expectation and determination you show will pay you and your family back in multiples at some point in the future.

Of course, things can and sometimes do go wrong, but if you master the following you're starting from a position of strength that puts you far ahead of the pack.

- Build a strong infrastructure for your company as described in this book.
- Understand the type of construction you will pursue.
- Know how to put a bid together.
- Identify the proper market for your business.
- Have access to adequate operating capital.
- Know how to manage money.
- Possess good judgment and managerial skills.

Starting a new business, particularly in the construction industry, involves significant risk due to factors including market fluctuations, regulatory requirements, and financial management challenges. If you are just starting out on your own, be prepared to take calculated risks and face potential setbacks with resilience and determination.

Successful entrepreneurs have a clear vision for their company, which includes setting, monitoring, and adjusting strategic goals as circumstances change. This vision drives decision-making and inspires confidence in your customers, financial professionals, and your employees.

As you know from your exposure to the construction business in some earlier capacity, starting and running a construction company demands a lot of hard work. From long hours to the hands-on management of projects, a strong work ethic ensures that you are prepared mentally and physically to handle the sometimes-extreme demands of the business, meet deadlines, and exceed customer expectations. My daily runs as a recreational runner during any particularly stressful period gave me an hour or so of sanity-preserving peace.

Comprehensive knowledge of the construction business is crucial. This includes understanding different building methods, materials, safety regulations, and staying updated on industry trends. Such knowledge is necessary for delivering quality work and also in making good business decisions. Set aside time to read construction-focused management books and magazines for helpful information and knowledge. Whatever sector of construction you pursue, there are likely numerous periodical publications that focus on your field from which you can get useful ideas. But also read general business and money magazines, such as *Entrepreneur, Forbes, The Economist*, and *Money*, to name a few.

An understanding of finance is essential for managing budgets, arranging for funding, pricing projects, and maintaining profitability. Financial knowledge also helps you manage cash flow effectively, critical for the survival of any new business. Learn how to analyze your company's financial statements. They shine a probing light on your past performance and suggest opportunities as well as guardrails for the road ahead.

The construction industry can be unpredictable due to, for example, changing economic conditions, weather, and client dynamics. I have described elsewhere in this book a period in which some of these conditions severely rocked my company. Successful entrepreneurs adjust strategies, pivot when necessary, and remain alert to new opportunities as they arise.

6 | 1 Entrepreneurial Characteristics

Challenging situations are common in construction projects. Finely hone your problem-solving skills to address and resolve issues to keep your projects on track, even at times when you do not have all the necessary information.

Quality in construction heavily depends on attention to detail. Your and your managers' conscientious oversight of projects is essential to reduce opportunities for mistakes, meet safety standards, and satisfy a demanding customer's requirements.

Adopt new technologies and seek solutions to construction challenges to give your new company advantages over your established competitors.

It is the rare person who possesses all of these entrepreneurial characteristics. Hone those you possess and work toward mastering others. They will serve you well as you navigate the complex landscape of starting and running a construction company.

Your Dual Role as Owner and Manager of Your Company

As the owner and the key element in the infrastructure of your construction firm, you fulfill two separate but equally essential roles: that of owner and that of manager/chief executive officer (CEO).

As the owner, you establish the business' objectives and vision, and you require yourself as CEO to carry them out.

In your role as the CEO, it is your responsibility to carry out these objectives by organizing, planning, controlling, directing, and facilitating the necessary elements.

Periodically throughout the year, step back and take a hard look at your performance as CEO against the predetermined business objectives you established as owner. The outcomes might meet your expectations, or they could fall short. If it's the latter, you as CEO must delve into and analyze the reasons behind it and detail them in a report. Then you the owner need to digest that report and determine the necessary internal changes that you the CEO must now put into place.

Managing the intertwined roles of owner and manager comes with demands. Fulfilling each role independently forces you to acknowledge any deviation from expectations and take action to get back on track. Without this continual performance check, you may default to simply comparing the current year's results against those of the prior year. It is easy to see how unhealthy that is.

Occasional deviation from your strategic plans and goals is to be expected, but if you maintain them as your benchmark, your home plate, you will always know where you are relative to them. Doing this reality check frequently should make it easy for you to see what went wrong and what you need to do to quickly get back on track. Do not allow the complacency that sometimes accompanies success to derail you from this foundational policy.

Things never stay the same. If they get better, it is usually because of good management practices. Without early attention and response to problems as they arise, they will get worse. As noted elsewhere in this book, one of the primary causes for the high five-year startup failure rate is the lack of knowledgeable and effective management.

Passion for Your New Baby Is Good, but...

I will venture a guess that you are a mature, rational, levelheaded man or woman as you enter your new life venture. And your ideas about starting a new construction business would not have gone beyond your initial thoughts without your passion and your drive.

But passion is a double-edged sword. Here are some tips that can help you avoid a few of the passion-driven pitfalls that can doom a business.

- Establish standard operating procedures as early as possible. This will help you avoid unproductive effort, time, and the frustration that goes with it.
- Delegate. Hire smart and responsible people and empower them to make decisions within the parameters you establish. Delegating is hard and I am good at telling you how important it is, but it took me too long to begin doing it.
- Set a schedule for yourself. Exceptions are inevitable, but it helps to have a plan that will allow for family time, recreation, and rejuvenation. Otherwise, you may lose the ability to know the difference.
- Establish an exercise routine for its immediate and long-term life-giving benefits. At least half an hour or hour per day.
- Consider developing a sounding board. Choose someone apart from your business you are comfortable opening up to. Their job is to listen, not to tell you how to run your business. This might be your wife or husband, a fellow businessperson, or your wise brother or friend. The pros and cons of a more structured advisory board are discussed later.

- Don't let your company become your sole reason for existence, which can result in burnout and other problems.
- Allow your mind to run free some of the time. Give it free rein. You may be surprised at how solutions to problems pop up in your head.

2

Company Culture

Your culture is your brand.

Tony Hsieh, former CEO of Zappos

A casual definition of a business' culture lies in the answer to the question, *What's it like to work around here?*

In more formal terms, your construction company's culture is the collective values, beliefs, attitudes, patterns, and behaviors that describe the shared identity and character of your organization. It's how your employees interact with each other. It sets the tone for the organization. It is its personality.

Peter Drucker, the management consultant, educator, and author, says *Culture eats strategy for breakfast*, emphasizing the idea that the organizational culture is more influential than even strategic planning in determining a company's success. That is a strong endorsement about the role of culture in an organization.

Your company's culture quietly manifests itself in the fundamental principles that guide your operations and decision-making processes. It is present in how your people:

- communicate,
- collaborate,
- handle conflicts, and
- share expectations regarding work ethic, accountability, and responsibility.

Information flow within your company is a function of culture as seen in how employees communicate with each other, in what way, and how often. Company culture is integrated into how you and the other leaders use your authority, provide direction, and cultivate the work environment.

Mastering the Construction Startup: A Business Infrastructure Guide, First Edition. Nick B. Ganaway.
© 2025 John Wiley & Sons, Inc. Published 2025 by John Wiley & Sons, Inc.

2 Company Culture

Culture grows from Day 1. If you want it to represent your and your company's image as you'd like to your customers, to your subcontractors, to your professionals, to the prospective employees looking for the right fit, to your visitors and vendors, and to your valued employees, who take pride in a work environment and structure in keeping with the nature of the business they're in – if these things are important to you – it will be you who guides them.

Intangible values such as the comfort, convenience, and appearance of your place of business and surroundings, your main office, and your jobsite trailer influence how others perceive your company and how they approach a business relationship with you, whether consciously or subconsciously. No construction site office or trailer is going to be perfectly neat and always clean, but your jobsite superintendent should set a standard for organization and cleanliness appropriate for a field operation.

If you want your employees to know they can express an opinion different than yours about a matter under discussion, or to follow a certain level of dress, or to be in the office at a specified time, or to interact with your customers in a certain way, or to maintain your policy against alcohol or drugs, or any other of your preferences, be assured these things will not come to pass unless you establish guidelines, both by your own example, in writing and through continual nourishment.

A high-profile example of the crucial role that cultures play in big business is the marriage and breakup of Daimler-Benz and Chrysler in 2007. Cultural differences between the exacting German engineering and corporate practices, and Chrysler's more informal American style are said to have led to significant internal conflicts. Knowledgeable observers believed their contrasting cultures were a major factor in the separation of the two companies that had merged in 1998.

Prioritize the culture and values that you want to define your company. They should be present in your company meetings, in your stationery, and in your conversational interactions with your employees. Employees who are worth retaining want to know what you stand for, what you believe in, and to some degree they will integrate these things into their work. Values and culture must not be an incidental part of your company but instead continuously made part of its being. It starts with you, its leader, but it becomes contagious and permeates the atmosphere. It will be recognized and rewarded by your customers, your vendors, and your outside professionals.

3

Elements of Leadership

Leaders are made, they are not born. They are made by hard effort, which is the price which all of us must pay to achieve any goal that is worthwhile.
Vince Lombardi – Legendary former
coach of the Green Bay Packers
professional football team

Leadership is a combination of skills, behaviors, and attitudes that can guide a person or group toward achieving a common goal. It's not as much about getting people to merely carry out directives as it is about inspiring them, promoting innovation, and guiding the organization through the bumps in the road that every organization encounters. As an entrepreneur and developing businessperson, you can enhance your effectiveness and impact by engendering the elements of leadership that you may have not yet accomplished. This chapter looks into these critical components, exploring both the personal and professional qualities that define successful leadership.

There is an age-old debate: Are leaders born, or are they created through education, training, experiences, and their recognition of needs and opportunities? In either case, accomplished leaders are known by the following traits and characteristics.

Vision and Strategic Direction

Leadership begins with vision that provides a clear direction for your organization and strategic planning. It's the leader's job to communicate his vision in a way that is both inspiring and relatable to all involved. This vision must align with the organization's long-term goals and also resonate on a personal level with employees, inspiring them to work toward a common future.

Mastering the Construction Startup: A Business Infrastructure Guide, First Edition. Nick B. Ganaway.
© 2025 John Wiley & Sons, Inc. Published 2025 by John Wiley & Sons, Inc.

3 Elements of Leadership

However, vision alone is not enough. Leaders also develop and implement strategies that guide the organization toward its goals. This involves understanding the market, recognizing opportunities, and managing risks.

Emotional Intelligence

A key aspect of leadership is emotional intelligence (EI). The essence of EI is understanding and managing your own emotions, as well as dealing with the emotions of others with understanding. Leaders with high EI are adept at encouraging a positive work environment, handling conflicts effectively, and maintaining morale even during tough times. It is in your personal and professional interest to understand the elements of EI and make them part of your being. There are many excellent books on EI.

EI is composed of several core skills:

- Self-awareness: Recognizing one's own emotions and their impact on others.
- Self-regulation: Managing emotions constructively, maintaining control and flexibility.
- Motivation: Employing emotions to pursue goals with energy and persistence.
- Empathy: Understanding, recognizing, and considering other people's feelings, especially when making decisions.
- Social skills: Managing relationships to move people in desired directions, whether in negotiating, leading a team, or communicating effectively.

Communication Skills

In my senior year in college, the instructor leading a management discussion brought up the importance for engineers to develop effective communication and writing skills. His point being that we might have created something as profound as, say, artificial intelligence, but if we cannot clearly and convincingly communicate its value to others beginning in its early stages, our valuable creation may never become recognized. Effective communication is the foundation of successful leadership. It's not just about conveying information but about dialogue, clarity, and an environment where ideas are weighed and exchanged. Successful leaders excel in various forms of communication including verbal and nonverbal cues. They are able to tailor their communication style to suit different audiences and contexts.

Adaptability and Resilience

The only constant in business is change. Leaders must therefore be adaptable and adjust course as markets evolve and new information becomes available. This paves the way to quickly respond to opportunities and effectively deal with threats.

Resilience is the ability to bounce back from setbacks. It requires maintaining a sense of purpose and optimism even in the face of failures. Resilient leaders inspire confidence in their teams and navigate through periods of uncertainty without losing momentum.

Integrity and Ethical Leadership

Leaders set the ethical tone for the organization. Integrity is fundamental, encompassing honesty, fairness, and consistency in decision-making processes. Ethical leaders inspire trust and loyalty, honor their commitments, and act in a socially responsible manner. Wise leaders prohibit any outward appearance of political bias.

Team Development and Empowerment

Leaders understand the importance of building and maintaining a strong team. This includes recognizing individual talents, delegating effectively, and empowering team members to take initiatives and make decisions. By developing a culture of trust and respect, leaders encourage accountability and enhance team performance.

Empowerment also ties into providing opportunities for professional growth and development. Leaders should mentor their team, providing feedback and challenging them to expand their skills and take on new responsibilities.

In summary, leadership is not a deluge but a stream of ongoing development and fine-tuning. It is continuous, evolving, shaped by experiences, challenges, and the perpetual goal of improvement.

Don't worry now about your opportunities or your effectiveness as a leader if you don't have full mastery of one or more of the above leadership characteristics. What leader would? If you fall short of where you want to be in one or more leadership characteristics, your strength in others will fill in for it.

Part 2

Regulatory Matters

4

Business and Government Regulations

I think it's very important that we have a regulatory system that is both efficient and effective. Good regulation is good for business because it promotes fair competition and ensures consumer trust.

Warren Buffett, CEO of Berkshire Hathaway

Incorporating Your Company

Incorporation requires filing Articles of Organization or a similarly named document with the Secretary of State in the state in which your company will be incorporated, its home state. The Articles of Organization will include your company's basic information, and you will be required to renew its registration annually and pay certain fees.

Registration in Other States

If your company will do business outside its home state, you will register it as a *foreign* company with the Secretary of State of each other state. This enables the state to know what companies or individuals are operating within its borders, impose its laws and regulations, collect fees, require the filing of forms, and collect taxes in states where applicable. Registration requirements vary from state to state.

Failure to register your business with a state can trigger severe fines, penalties, and other consequences including being barred from the use of the state's court system. For example, if you were to do business in that state before duly registering your business and subsequently needed to sue an entity for nonpayment,

---

Mastering the Construction Startup: A Business Infrastructure Guide, First Edition. Nick B. Ganaway.
© 2025 John Wiley & Sons, Inc. Published 2025 by John Wiley & Sons, Inc.

4 Business and Government Regulations

liability, or other reasons, you may not be permitted to use the state's courts to do so, clearly a severe consequence. Some states that require a professional license, e.g. a general contractor license, impose the same penalty for failing to obtain a license as required by state law, prior to beginning work.

Registered Agent

You'll have to name a registered agent for each state you register to do business in, which can be a person or an organization. A purpose of the registered agent is to accept legal service as to any official notices or lawsuits against your organization and deliver them to you or another designated person in your company for filing an answer. Timing is critical in some situations. Failure to respond to a lawsuit within the specified period can result in a default judgment by the court awarding the plaintiff all requested benefits. So choose a responsible party as your agent. You may choose a lawyer or one of the many firms that provide this service, such as ZenBusiness or LegalZoom.

Professional Licensing

If a state requires a general contractor's license, you must apply to the designated board or agency, comply with its rules and regulations, and in some states or jurisdictions pass a written examination and provide financial information. The severe penalties described earlier for conducting business before obtaining the required license(s) may apply. There may also be certain insurance requirements for registration.

If your company is performing services for a municipality or other governmental entity, you may be required to provide a bond to guarantee you will meet certain requirements. Your insurance agency will be able to help with this. Performance and payment bonds are discussed in Chapter 14.

Municipal Business Licensing and Permitting

In addition to the aforementioned requirements, you will be required to obtain a business license in the city or county (called parish in Louisiana) where you do business. This is typically renewable annually.

Professional licensing as described earlier may also be required at the city or county level.

Depending on your type of business, a specific permit may be required for a given project.

Corporations and LLCs

There are several types of legal business formation, and you need to understand their different requirements and applications. The most commonly used small-business entities are described here (Table 4.1).

A limited liability company, or LLC, is considered a corporation if the LLC owners elect it to be treated as a C or S corporation for taxation purposes. However, for purposes of this discussion and throughout this book, it is assumed that you as owner of your LLC will not have made that choice and therefore your LLC is not considered to be a corporation, except when otherwise identified.

There are important differences and likenesses in LLCs and corporations.

Your business entity type determines how your company is taxed. Corporations and LLCs are legal entities that stand alone and independent from their owners, who generally benefit from protection from claims and lawsuits against those entities.

The US Internal Revenue Service (IRS) allows corporations and LLCs to hold a limited amount of money for working capital, meaning it is shielded from taxation, a big advantage to you. The allowed amount is nonspecific, varying depending on the nature of the business. You and your certified public

Table 4.1 How US businesses are legally organized.

Type of Business	Non-employer	Small Employer	Large Employer
Sole proprietorship	86.6%	13.7%	5.7%
Partnership	7.4%	11.9%	25.2%
S corporation	4.5%	52.1%	30.9%
C corporation and other	1.5%	22.5%	76.2%

Source: Small Business Administration Office of Advocacy / Public Domain.

4 Business and Government Regulations

accountant (CPA) should make the case justifying the largest reserved amount the IRS will allow.

If your corporation or LLC cannot pay its debts, its creditors can claim its assets but not your personal assets. As the owner of a corporation or LLC, you generally risk only the amount of money you invested in the business plus any debt you personally guaranteed. You must meet certain criteria to maintain this protection, including maintaining strict separation between your business and personal accounts and other personal assets.

If your business is just starting up, you will likely be asked by your creditors to personally guarantee payment of any debt incurred by your company. Think very carefully before agreeing to such a personal guarantee, as it exposes your every personal asset to risk. Explore options such as a fixed-dollar amount guarantee or the shielding of certain of your assets from any required personal guarantee.

Even though you have the intended protections afforded by your corporate or LLC status, it is important to obtain liability insurance that will cover both you and your company in case of a lawsuit or other claim.

If you have substantial personal assets, consider also obtaining an umbrella insurance policy that will protect you against claims that exceed your primary policies. In addition, you may also want to consult a bankruptcy or estate attorney to seek ways to further insulate yourself personally from your company's business obligations. This is a matter of prudence, not an indication that you doubt your businesses' viability.

Your personal assets may lose corporate or LLC shielding from legal liability if you comingle your business assets with your personal assets. Lawyers refer to this as *piercing the corporate veil*. To avoid this, keep business records, bank accounts, and finances cleanly separated from your personal finances and assets. The company must have its own bank accounts and forms of credit. Purchase orders, trade accounts, contracts, agreements, and other business documents should bear the company name. Sign all business transactions and documents in the name of the corporation, e.g. Smith Manufacturing Corporation., by William Smith, President. In the case of LLCs, sign it as Smith Manufacturing Co., LLC, by William Smith, Owner (or Manager).

If your corporation or LLC is sued for any reason, plaintiffs' lawyers often pursue any entity remotely associated. For example, if your LLC or corporation is sued for causing damage to another party, the plaintiff's attorneys may file suit against you personally. But if you have maintained strict separation between your company and yourself as an individual, you are likely to be protected. Note that such protection is not the case with sole proprietorships or unincorporated partnerships, described below.

If you own more than one company, don't lump them into a single LLC or corporation, but instead set up each one as a separate LLC or corporation. Then, if one of your companies is sued, only that specific company's assets are at risk, while your other companies remain separate and apart. Each LLC or corporation must have its own unique name.

Don't keep excess money and unnecessary assets in a corporation or LLC. If it is sued, those assets are at risk. Regularly monitor and adjust the assets held in each of them.

The LLC

Upon exploring the various legal entities, you are likely to find that the LLC, as distinguished from a corporation, meets your needs and offers you the most flexibility and simplicity. Here are some of the LLC's advantages.

The LLC is not subject to income taxation. It is a "pass-through" entity for tax purposes. The income of an LLC flows through to the sole owner, who reports it on his or her personal income tax return, thus the pass-through label. The LLC avoids double taxation inherent in the "C" corporation.

The regulation of LLCs varies from state to state, and some states use a title other than LLC. LLCs have fewer regulatory requirements than the C corporation (C Corp), and a board of directors is not required. Various required filings and fees are ongoing but are not necessarily burdensome.

The LLC in most jurisdictions requires a formal *operating agreement*, which describes the roles of the members, the management structure, financial arrangements, and other operational details. An operating agreement is similar to a partnership agreement in purpose. It is usually written by a lawyer.

This book is written with the LLC form of business in mind unless stated otherwise.

The S Corporation

An S Corp may be recommended or required in some cases. S Corps are similar to LLCs. Here are a few of the differences.

- An S Corp passes its profits and losses to its shareholders for federal tax purposes, thus avoiding the double taxation inherent in the C Corporation.
- S Corps are limited to 100 shareholders.
- S Corps cannot have non-US citizens as shareholders.
- If you plan to be a high-growth company and raise capital, an S Corp may be preferable to an LLC.
- An S Corp requires more filings, reporting, recordkeeping, and other duties than the LLC.
- An annual shareholder meeting is required and must be documented in the minutes.
- A board of directors is required. If you are the only shareholder, you can be the sole director and also hold the official positions.

The C Corporation

- You may be a single shareholder in a C Corp, but in many circumstances that is like using a sledgehammer to drive a nail.
- The C Corp is cumbersome and relatively expensive to maintain.
- The C Corp is required to file its own tax return, have a board of directors, and maintain formal meeting minutes.
- C Corp profits can effectively be taxed twice – first when the company reports a profit, then again when dividends are paid to shareholder(s) and reported on their personal tax returns.

Corporations are required to have bylaws, which lay out the roles and responsibilities of directors and officers, how decisions are made, and the process for holding meetings.

Corporations also require Articles of Incorporation to be filed with the state to legally establish the corporation and describe its structure, including its name, purpose, and stock particulars.

LLC operating agreements and corporation bylaws are similar in that each is crucial to ensure clear governance and protection for the entity's legal status.

My attorneys have cautioned that complying with the formalities required for corporations and LLCs is necessary in order to defend against lawsuits attempting to circumvent their owner's sheltered legal status.

Sole Proprietorship

A business you own exclusively in your name or in a name you assign to it is a sole proprietorship. There are several advantages to this business structure, making it tempting to the startup.

- Easy setup.
- You are the boss. Period.
- Ongoing relative freedom from government regulation.
- Low organizational cost.
- It's not a formal structure, so incorporation is not required, and typically there are no filings or paperwork to be completed before you get started. In some cases you may need to get a business license.
- In contrast to the several relatively burdensome annual filings required of corporations and even LLCs, a sole proprietor is not faced with annual reports or filings with the state.
- All profits or losses are filed on your personal tax returns, minimizing additional cost for business accounting or bookkeeping services.
- Recordkeeping for a sole proprietorship can be as simple or sophisticated as you want. However, for purposes of organization and analysis, a record-keeping system that maintains separation of business and personal finances is recommended.

I emphasize here that despite the advantages available for the sole proprietorship, there is a huge potential price to pay for this simplicity: Your assets including your home, automobiles, bank accounts, investments, and other personal property are at risk. A sole proprietor is personally liable for any debt that arises out of lawsuits, claims, or business failure.

Partnership

It is common for two or more people sharing a mutual business interest to enter into a partnership. The terms of agreement by which it is formed should be carefully considered with the help of a lawyer. While partnerships are usually friendly and cooperative in the beginning, they can become untenable due to disagreements, changes in a partner's personal circumstances, or other reasons.

When this happens, the departing partner very likely wants to leave with his share of the value of the business, but the remaining partner may not have the cash or even the willingness to buy him out. Also, it can be difficult to accurately determine the value of the business without a costly and delaying appraisal and the services of an accountant. Appraisals are to some degree subjective and open to dispute. The remaining partner in a construction business may want to restrict the departing partner from setting up shop as a competitive contractor; this restriction is often done by way of a difficult-to-enforce noncompete agreement (discussed in Chapter 12) and a confidentiality agreement.

Also, be aware that if one or more partners are unable to bear their share of the financial load in case of claims, lawsuits, business loans, business failure, or other liabilities, the remaining partner(s) automatically assume the burden for the full amount, jointly or individually. This means that if there are multiple partners but only one of them is financially able to cover a loss, that partner becomes liable for the entire amount.

Another situation arises if one of the partners simply wants to sell his or her share of the partnership to an outside party. Without an agreement in place to prevent it, a partner can sell his share to a third party – *any* third party. The remaining partner then will have to deal with his new partner, who may be totally unsuitable for the remaining partner for any number of reasons.

Partnership Agreement

The scenarios described earlier can be addressed by way of a *partnership agreement* between all of the parties, which describes how the partnership's management and ownership are structured and spells out procedures for partnership breakup and other circumstances like the ones mentioned earlier if they arise. This requires a lawyer, and you should consider hiring your own to represent your interests rather than one who represents both or all partners. Any two or more people considering forming a partnership put themselves at undue risk if they do not enter into a formal partnership agreement.

The italicized caution stated earlier for single proprietorships applies equally to partnerships. I hope I have sufficiently dissuaded you from doing business as a sole proprietorship or partnership without LLC or corporate protection.

Establishing Your Corporate Entity

Do not make this critical decision without professional help. Your CPA will advise you on type of entity and tax considerations, which vary based on type of entity. Your attorney will help you in establishing your company as required by law and other regulations and to prepare the associated legal documents including the operating or partnership agreement.

You may find relatively inexpensive forms and documents at Legalzoom. com and other online law firms that you can use in some instances, but you forego legal advice tailored to your unique circumstances by your own lawyer.

See Appendix A for a Sample Partnership Agreement.

Choosing the Name for Your Company

Consider the following when choosing the name for your business.

- Ease of spelling and pronunciation.
- The company name as registered with your home state must be unique. If you plan to operate outside of your home state under the same business name, the name must not already exist in that state or states.
- Decide whether you want your business name to suggest what products or services you provide, e.g. Thompson Contracting Company or The Thompson Company.
- Keep in mind that if two or more people forming a business include multiple partners' names in the company name, this could become inconvenient or worse in case of a company split or if a partner drops out.

Part 3

Getting It Done

5

The General Contractor

The construction industry depends on skilled general contractors who can orchestrate diverse trades and bring complex projects to life efficiently and safely.

Doug Oberhelman, Former CEO of Caterpillar, Inc.

As the general contractor, you are the hub of the construction project, coordinating and managing the entire process to completion. Your duties encompass a wide array of responsibilities, starting from the initial planning stages to the completion of the project.

Making the decision to become a general contractor should not be taken without identifying, understanding, and committing to the responsibilities that come with that decision. While Chapter 1 discusses the broad spectrum of necessary requirements of the typical entrepreneur, those of a general contractor go further.

It is likely that most people who decide to become a general contractor cut their teeth as an employee of a general where they rose through the ranks. But there is no prior experience that brings them face to face with the realities encountered by the owner and operator of a general contracting firm, i.e. *the general contractor.* With every decision the general contractors make, they have at stake their financial, social, and personal future, that of their family, employees, and to some extent their subcontractors.

The general contractor develops a comprehensive timeline and an accurate budget to meet the requirements of a given project in accordance with the plans and specifications and the conditions under which the project will be built. This includes the location, access, level of security needed, staging and parking area, soil condition, typical weather, workforce availability, impact of street traffic, available utilities, potential interference with the project by existing utilities,

Mastering the Construction Startup: A Business Infrastructure Guide, First Edition. Nick B. Ganaway.
© 2025 John Wiley & Sons, Inc. Published 2025 by John Wiley & Sons, Inc.

5 The General Contractor

proximity to other structures, and other site-specific considerations. The general contractor manages any changes or modifications to the original plans and specifications and coordinates with the project owner and other parties to ensure that adjustments are integrated without compromising the integrity of the project.

As a startup general contractor, you may fill all the key jobs yourself at first – including preparing bids, managing projects, hiring subcontractors, and procuring materials.

All too often, intelligent men and women start their own businesses as a general contractor only to find out later, sometimes with severe consequences, that their experience, knowledge, financing, or maturity of judgment does not sufficiently qualify them. Fortunately, none of those factors is permanent. Many very successful contractors, as well as entrepreneurs in other fields, became successful only after once finding themselves in such circumstances.

The Owner–Contractor Relationship

The relationship between the owner of a commercial project and the general contractor is a complex and dynamic partnership grounded in mutual trust and clearly defined responsibilities. This relationship is crucial to the successful completion of a project, as it encompasses the planning, execution, and completion phases, each marked by distinct duties and obligations.

At the outset of a commercial project, the owner, who is often a business entity or an individual investor, embarks on the planning phase with specific goals and objectives. The owner's primary responsibility to himself and to all stakeholders, including the general contractors who submit bids for the project and one who ultimately builds it, is to provide a clear vision of the project, including its scope, budget, and timeline. This involves assembling a team of architects and designers to develop detailed plans and specifications. Once these plans are in place, the owner solicits bids from various general contractors, selecting one based on factors such as experience, reputation, and cost.

Upon selection, the general contractor enters into a formal contract with the owner. This contract, usually referred to as a construction contract or agreement, delineates the terms and conditions governing their relationship. It outlines the project's scope, schedule, budget, and the specific duties each party must fulfill.

The general contractor's primary duty is to execute the project according to the plans and specifications provided by the owner. This includes managing subcontractors, procuring materials, and ensuring that all work adheres to industry standards and regulatory requirements.

During the execution phase, communication and coordination between the owner and the general contractor are vital. Regular meetings are held to review progress, address any issues, and make necessary adjustments. The general contractor provides the owner with updates on the project's status, including any potential delays or budget overruns. It is the contractor's duty to proactively identify and manage risks, ensuring that the project stays on track.

The owner, on the other hand, is responsible for providing timely payments to the general contractor as stipulated in the contract. These payments may be made in installments, typically monthly, based on the completion of specific project milestones. The owner must also make prompt decisions on any changes or modifications to the project (change orders) that may impact the scope, cost, and schedule. Effective decision-making by the owner is essential to avoid delays and additional costs.

One of the critical aspects of this relationship is quality control. The general contractor is responsible for ensuring that all work meets the required standards and specifications. This involves conducting regular inspections and addressing any deficiencies promptly. The owner, often through a representative or project manager, also conducts inspections to verify that the work is being carried out as agreed. Any discrepancies or issues identified must be resolved collaboratively to maintain the project's integrity and quality.

As the project nears completion, the general contractor's duties shift toward finalizing the work and preparing for project handover. This includes completing any remaining tasks, conducting final inspections, and addressing punch list items – minor defects or unfinished work identified during the final walkthrough. The owner is responsible for accepting the completed project after ensuring that it meets his expectations and the contractual requirements.

Upon successful completion and handover, the general contractor provides the owner with all necessary documentation, including warranties, operation manuals, and as-built drawings. This marks the end of the general contractor's direct involvement in the project, though he or she remains responsible for addressing any post-completion issues that arise during the warranty period.

In summary, the relationship between the owner of a commercial project and the general contractor is built on a foundation of mutual trust, clear communication, and well-defined responsibilities. Both parties must work collaboratively to navigate the complexities of the construction process, ensuring that the project is completed on time, within budget, and to the highest standards of quality.

The remainder of this chapter and those following bring focus to the role of the general contractor.

Managing Risk

Management of risk is fundamental to every aspect of running a successful construction company. The construction industry, perhaps more than many others, is burdened by uncertainties that can significantly impact your project's timeline, budget, project owner satisfaction, and overall success. Understanding and preparing for them can lessen their effects and help maintain the stability and profitability of a given project and of your company. Following are several critical risk factors that construction companies must manage to ensure effective operations and project completion.

You or your bankers or attorney should develop financial and other pertinent information about the project owner prior to your investing time and resources in the project.

Implement an effective estimating process with experienced estimators and a historical databank. Leveraging technology, such as construction estimating software, can also reduce the risk of costly errors in the bidding phase.

Begin the project with clear payment terms agreed to contractually by the owner with specific remedies available to you. Regular clear communication with the owner about job progress and any glitches helps ensure timely payments. Establish and maintain your company's cash reserves as early as possible as a buffer against unexpected payment delays and other cash flow situations. Establish a line of credit with your bank for this purpose. But do not think of this line of credit as your working capital. Keep it separate and pay it off as early as possible after the need for it subsides.

The construction industry is also vulnerable to external factors such as bad weather. While weather conditions are beyond your control, build in the assumption that they are going to affect your outside projects and incorporate weather contingencies into project schedules and budgets.

Managing Risk | 33

Nonperforming subcontractors represent another significant risk. Subcontractor failure almost certainly leads to project delay, quality issues, and increased cost. This underlines how crucial it is to thoroughly vet subcontractors by checking their credentials, financial stability, past performance, and reliability. Building strong relationships with reliable subcontractors and having a backup plan in place can also provide a safety net if problems arise. All of these factors are discussed in Chapter 6.

Another risk factor is the problem of reducing overhead fast enough when work slows down. Reducing payroll is an obvious way to cut costs almost immediately but can have long-lasting repercussions such as loss of key employees, lowered morale, and reduced readiness when things return to normal. Regularly review your overhead expenses and identify areas for cost savings, which will enhance your company's ability to adapt to changing market conditions and maintain profitability.

Disputed change orders are another common risk in construction projects. If not managed properly, change orders lead to disputes, project delays, and increased costs. To degrade this risk, establish a change-order process including documentation, approvals, and communication procedures. Effective project management and regular updates can help manage change orders efficiently, keeping projects on track and within budget.

Delayed payments by project owners pose a significant cash flow challenge. This can stem from owner–contractor disputes, owner cash flow problems, contractor delay in submitting required documentation to the owner, deliberate or inadvertent slow payment by the owner, and other reasons. Delay in payment from the owner can stretch your resources to the limit, affecting your company's ability to pay subcontractors, suppliers, and employees promptly.

Maintain strong contractual agreements and monitor the financial health of the project owner throughout the project. If you allow a late payment or other breach of your agreement, notify the project owner that this is an exception and not a modification of the contractual provisions. If it continues, consider giving the owner official notice as provided in the contract, considering all factors related. Remember that allowing the breach to routinely continue will make it harder to stop.

Slow delivery of construction materials can delay project timelines and increase costs. Develop strong relationships with reliable suppliers and have contingency plans for alternative sources in place. Monitor supply chain trends and maintain open communication with suppliers to ensure timely deliveries and address any potential issues promptly.

34 | *5 The General Contractor*

Your loss of a key employee is another risk factor. Cross-training of employee duties and responsibilities among key players can, to some degree, reduce this impact. Effective employee management practices discussed elsewhere in this book help to ensure that the company remains adaptable to personnel changes.

Construction Type

You must identify and evaluate the size and category of construction you will follow. Startup contractors often stay with the type of construction they learned from experience, but others might try light commercial, industrial, retail, government projects (of which there are myriad types) or others.

Specialization in a certain niche of construction can be rewarding and is described in Chapter 16. After deciding what type and dollar-size projects and location you want to pursue, you will be able to tailor your company to fit, which is imperative for a startup firm whose resources may not support a wide range of dissimilar project types, sizes, and distant locations.

Union or Nonunion

Avoiding union work is my preference, but this may not be practical depending on where you operate. If that is the case, be sure you understand the local union rules and take them into account when working up a bid.

As a general contractor, you appreciate the level of independence that comes with being your own boss and making your own decisions, but on union projects the union becomes your business partner. Your autonomy and flexibility are substantially reduced.

Your cost will go up because the higher wages you pay are multiplied by the additional hours that result from union rules and slowed progress. Mandatory conditions such as work hours and breaks limit the contractor's flexibility. Nonunion contracts typically get completed in less time and for less cost, apples to apples, than union projects.

Union work rules are Project Labor Agreements (PLAs) required for unionized projects and increase the opportunity for legal disputes to arise.

Work stoppages, which may take place during a union dispute, cause the added problems of interrupting the rhythm of the ongoing construction operation, getting subcontractor crews back on schedule afterward, and rescheduling materials and equipment deliveries.

If your projects are in right-to-work states, you may be able to fend off your employees' receptiveness to unionization by examining and tweaking as necessary your openness and transparency, opportunities for additional education, pay, and employee benefits. (See more about what employees want in Chapter 12.)

Managing the Common Causes of Contractor Failure

Thomas C. Schleifer, a former general contractor, writer, workout manager for troubled construction projects, and educator, had the opportunity to study specific cases of contractor failure over time to determine the primary causes and detailed them in his timeless book, *Construction Contractors' Survival Guide* (Wiley). The contractor who understands and conducts his business in accordance with Schleifer's research is infinitely better prepared to navigate the challenges that come along.

Schleifer notes that the following are prime causes of contractor failure.

Increase in Project Size

Growth of a construction company is normal, within limits. By increasing your typical job size *gradually*, you can manage the effects on increased project time, working capital, required level of supervision, cash flow, and introduction to various new methods and materials.

But when you are in need of a project to keep your crews busy, the pressure on you to land one increases and you begin to test the waters for business outside your usual realm. If you win a bid twice the size you've been doing, planning the project will be much more difficult. Calculating the additional resources that will be required, and acquiring them will likely be problematic. The longer job duration will result in increased retainage and a strain on your cash flow. The level of supervision as well as your own involvement will increase. Scrutiny by the owner, inspectors, and lenders will be more

36 | 5 The General Contractor

disruptive than you are used to. Schleifer says you can probably get the job done, but making a profit is a different challenge.

Changing Your Geographic Area

Circumstances, such as lack of work in your usual area or when special opportunities beckon elsewhere, often compel contractors to look for jobs in unfamiliar locations. Moving your operations to a new geographic area is fine, but you must consider the differences you will encounter. Even if you are doing a familiar kind of work, you may run into different regulations, methods, customs, and weather conditions, not to mention attracting subcontractors in a busy market. In addition, your labor situation will change even more significantly if you move from a nonunion area to where union work prevails.

Change in Key Personnel

Your company is no longer a proven organization if you lose an employee who manages one or more of the most foundational roles, i.e. estimating, construction operations, administration, or accounting. The longer this situation exists, the greater the risk to your company because the remaining functions to some degree receive less attention as their managers take on more responsibility.

Taking on a New Type of Construction

There can be any number of reasons that a contractor decides to move into a new type of construction. Before doing so, there is a need for research and planning, including recognition that you will pay a price for the adjustment required for your firm to learn to perform different types of work. Even hiring one or two senior people who are well experienced in the new type of work does not promise early success and you should prepare for loss of profit as you make the transition.

Lack of Management Maturity

Outgrowing your management capacity is often paired with one or more of the other elements of contractor failure. It can be hard for a growing organization to maintain the necessary level of qualified managers in key positions.

Such insight into causes for contractor failure has unlimited instructional value for those of us who are contractors and those who will become one.

In light of these common causes for failure, it is easy to see why the five-year failure rate of construction startup businesses is the highest among all kinds of enterprises. Certainly, luck and many other factors play a role in any successful venture, but to count on luck to offset the risk inherent in violating a combination or often even one of the elements of contractor failure described here is to invite serious setbacks if not outright failure.

However, having knowledge of these elements is worth nothing unless you incorporate them into your decision-making process. They will come into play in many of the circumstances you confront, as you seek every opportunity to profitably grow your business.

For an in-depth discussion of these causes for contract failure, refer to Schleifer's book. Understanding and working around them will be a key factor in your contracting success. The book has been my go-to reference for many years.

There is an episode in my history as a general contractor that presented several elements of contractor failure described earlier. After we had built a number of restaurants for the now-defunct upscale fast-food chain D'Lites, I received a phone call one morning from a man who had become a new D'Lites franchisee for parts of Pennsylvania. "Would I like to build his restaurants there?" he asked. I was interested in getting new business but expressed doubt. We worked in the Southeast, far from Pennsylvania. After some discussion, he said it would be negotiated work and he planned to open a number of units over the next few years.

Talk is cheap, but these potential terms were interesting enough that I flew up to meet with the franchisee to consider work, even knowing it would directly conflict with more than one of the conditions for failure just described earlier. Pennsylvania was a geographical giant leapfrog from my familiar area of operation, certainly not part of methodical expansion.

While talking with contractors in the Pennsylvania area, I learned that nonunion work was practically impossible. Meeting the franchisee's timetable and work volume would double my company's capacity in terms of manager-level employees, working capital, and my personal involvement. We were not used to severe weather conditions. Some of my people would be working far from their homes. Any of these circumstances should challenge any contractor's common sense. Yet banks and sureties tell us that these

5 The General Contractor

guidelines are often ignored, thereby upsetting or destroying the livelihood of contractors and their employees, subcontractors, and suppliers. Declining the Pennsylvania opportunity was an easy decision.

Task Dependency Software

For construction projects of any complexity, you can use task dependency software such as Microsoft Project. This software allows users to prioritize tasks (step-by-step elements of the process), deadlines, and dependencies to create a detailed project schedule.

It helps the manager visualize the project timeline so as to schedule various tasks when they can first be performed, for example, potentially eliminating lost time before the subsequent task is begun. It promotes collaboration among team members and other parties in the loop.

Task dependency software requires the user to have a deep understanding of the project and of construction processes. There is a learning curve for new users as they come to understand its features, functionalities, and terminology.

The software's output is of course dependent on the accuracy of the input. To be of value, the software must be continually updated as the timing of the tasks changes during the project.

Construction Document Software

Consider also construction document software such as Dropbox Business or Google Drive. Your lawyer's recommendation will be to *document, document, document* incidents because the outcome of construction lawsuits often favors the party who has the most, the best, and the most contemporaneous documentation. This includes text messages, emails, paper documents, and informal handwritten notes made in real time and preserved.

Construction Bookkeeping Software

Research software products designed for construction firms, which can make your bookkeeping more efficient.

Awareness of the Economy and Other Business Threats and Opportunities

Learn and follow the many factors that shape the economy. No one can predict with certainty when an economic recession will come along, but successful contractors keep their eyes wide open for signs that can indicate the coming of a downturn.

You can't control the economy or many other perilous factors but by conscientiously maintaining awareness you can be prepared for what may come. Watch for the periodic or triggered pronouncements by the US Federal Reserve for clues about interest rates and other economic predictors. An increase in interest rates affects developer decisions that usually trickle down to the construction industry. Stay in touch with your customers, who may tell you what they see ahead. Investigate the causes behind any changes in their usual business behavior for important clues. Maintain communications with your banker and companies in your business orbit. Read extensively construction industry magazines and newsletters and local and national newspapers for informed opinions and forecasts.

Keep an eye on weather forecasts. Few industries' operations are more dependent on weather conditions than the construction industry.

Be aware that bad things often happen in good times. Prosperous conditions in your company and in the overall economy sometimes lure business owners into making decisions that don't turn out well. In Chapter 18, I have discussed such an episode in my own business.

Negotiation

Construction contracting is hard-nosed, requiring tough calls on your part, and it helps if you are a successful poker player. Aside from face-to-face negotiation, putting the final markup on a proposal you're submitting is also tough, especially if you really need the job. Contractors who bid mostly on lowest-bidder projects are often tempted to shave their bid too much, which can lead to trouble.

In an interesting twist on bidding, we once submitted a bid and a week later the franchisee called and told me that the two lowest bids were separated by exactly *one dollar*. The franchisee, unbelieving that such a coincidence occurred purely by chance, was not willing to award the contract to either of us on that basis. Certainly

40 | *5 The General Contractor*

it seemed nearly impossible. What are the odds that two independent and honest bids on a million dollar project would be a dollar apart?

But if two contractors were colluding, what would be gained by agreeing on such a bidding strategy? I knew our bid was honest. Our policy was to be straightforward and besides that I was in the room when we decided on the final price and sealed the package. I learned later that we were the dollar-low bidder, but the franchisee re-bid the job, and we won it *fair and square* – again.

(Contractors who lose a bid by a large margin sometimes rationalize that the winning bid was low due to some omission or calculation error by the winning contractor and that he was certain to lose money on the project. There's an old joke about a contractor who went to the bid opening of a multimillion dollar project only to learn that the bid price of the winning bidder was a hundred dollars lower than his. The losing contractor angrily walked away from the architect's office shouting to no one, *"That fool is going to lose his butt."*)

Proactively Manage Budgets

You expect your project managers and every employee who is responsible for a budget item to continually monitor it and take corrective action when deviations occur. This feeds into your overview of the big picture.

Many business owners will be able to tell you of an instance when they were blind to a cash flow problem or when a business loss caught them off guard. As will many *former* business owners, whose unawareness of their financial status led to them going broke. By meeting regularly with your project managers and accountant, listening to their reports, and asking probing questions, you reduce the likelihood of learning of a problem too late. Watch for vague answers to your questions for no one wants to give you bad news. Don't stop asking questions until you get the answer.

Budget control starts with close monitoring by each project manager, jobsite superintendent, office manager, and anyone who has spending authority. Cost deviations from the internal budget must be accounted for and addressed whenever project cost is discussed.

An ambitious project manager may decide not to report any savings he has achieved through negotiations, change orders, etc., so that he can use such savings to offset cost overruns on other line items. This kind of cost manipulation by project managers or others prevents you, the contractor, from having reliable current cost data. This can affect decisions you make or leave you in the dark about a potentially more serious underlying problem and prevent you from taking necessary corrective action. You may be able to control this practice by telling your employees the serious potential consequences and that you expect the practice to stop.

Demand High Quality

There are many opportunities for things to go wrong on a construction project, but excellence is a valid goal. You do not want to be known as a contractor that produces *average* quality. Make it known within your organization that excellence is what you expect so that it will become a trademark of the company. If you offer excellence in your promo materials or your pitch to a project owner, that is what they expect.

Achieving excellence in all that you do for a customer begins with hiring excellent employees who are committed to that goal. It will become one of the characteristics that mirror your company's culture and leave an imprint on your customers' minds.

The definition of excellence depends on the type of construction. Excellence for structural concrete is different than excellence for medical offices.

Cultivate Relationships

Interpersonal relationships among your employees and theirs with you are the factors involved in forming your company's culture, which affects employee satisfaction with their job. Positive relationships lead to more productive and committed employees. Satisfied employees form a bond of trust among themselves and with company management. They are more likely to collaborate among the employee group to achieve organizational goals and with pride. Employee turnover rate has been shown to be directly tied to employees' overall satisfaction.

Just as it is with maintaining relationships among your employees, positive and effective cooperation between your people and your customers, your prospective customers, your vendors, your subcontractors, and your bankers altogether should be nourished.

If you've ever dialed up a Ritz-Carlton Hotel, I bet that the employee you spoke with made you feel welcome, that their time was your time, and that they were happy to be of service to you. You will feel the same if you visit the hotel. The person(s) answering your office phones often have the opportunity to create a caller's first impression of your company, which may set the tone for any further contact between the caller and the company. That initial contact experience is not likely to be forgotten.

You will not accomplish excellence on every project, but if you continually promote it in word and deed and if everyone on the team strives for it, excellence will become a hallmark of your company. A company that does not set standards for itself will not be aware when improvement is needed. Your customers will.

Meet the Schedule

Delays in the project schedule can impact the budget, while poor quality leads to increased costs and safety hazards. Similarly, safety lapses can result in delays and additional expenses.

Your company being known as a contractor who consistently completes projects on time might be an architect's or owner's deciding factor in their selection of contractors to include on their bid list or to negotiate work with. Owners may schedule delivery and installation of machinery, fixtures, and equipment with long lead times. They often hire employees based on a specific start date. Successful contractors meet their delivery dates.

Stakeholder Communication

Successful projects also require effective communication among all parties, including clients, contractors, subcontractors, suppliers, and regulatory bodies. Clear and regular communication helps in addressing concerns promptly and making informed decisions.

Human Resource Management

Ensuring that the workforce is well-trained, motivated, and efficiently managed is essential. Human resource management involves recruitment, training, performance management, and ensuring a positive work environment. Employee relations are discussed in Chapter 12.

Trust

The degree of trust and the quality of relationships among team members and between employees and management begin with you, the contractor, following through on your commitments to your employees and others in your business orbit. This is a hallmark of successful contractors in their internal and external business relationships.

Marketing

Creating an effective marketing program for your commercial construction firm requires a strategic approach to reach prospective customers.

Begin with a professional website to tell your company's story. This includes who you, the owner, are, your related history, and credentials that qualify you as a general contractor. Introduce your key employees. Explain the services your company provides and specializes in if that is the case. Your comments should avoid platitudes and clichés but effectively demonstrate your commitment to quality and the other characteristics you want to be known for. Secondary website pages can include client testimonials and case studies. Go strong on appropriately large high-quality (professionally made if possible) photos of completed projects for a lasting image in the viewer's mind. Include your web domain on your business cards and stationery.

Your brochures and flyers should incorporate much of the same and can be distributed at industry events and sent directly to potential clients.

You can produce short videos that highlight your completed projects, client testimonials, and behind-the-scenes looks at one or two of your projects in progress. Video content can be very engaging and effective in demonstrating your capabilities.

Secondary materials may also include challenges encountered, the solutions you provided, and the outcomes achieved. These can be shared on your website, in email campaigns, and in meetings with prospective clients.

Attend industry events and rent display spaces to attract and meet potential customers. Have your marketing materials ready to distribute and engage in conversations to build relationships. Neither you nor any of your employees, even your jobsite superintendents, should ever be without printed business cards bearing name and title in print. You never know when handing someone your card might make a difference.

5 The General Contractor

Get involved in community organizations. Participate in your local chambers of commerce and service organizations such as Rotary where the qualities you stand for will be seen by others who may someday directly or indirectly impact your business. Become a member of the Associated Builders and Contractors (ABC), a national, highly respected organization that is a strong advocate for the construction industry and individual contractors. The same can be said for the larger Associated General Contractors (AGC). Both are sources for industry information, materials, forms, and networking with fellow contractors.

You can put your name before potential customers by building an email list and sending out regular newsletters featuring your latest projects, industry news, and insights into the construction process.

Don't overlook social media marketing. You can use platforms like LinkedIn, Instagram, TikTok, Facebook, and others to make your company known. LinkedIn is particularly useful for B2B marketing, allowing you to connect with decision-makers in your target companies and industries.

Write articles, blog posts, etc., about trends in commercial construction, tips for successful project management, and the benefits of working with a professional construction firm. Share this content on your website and social media channels to establish yourself as an industry expert. I have written numerous articles for publication in several construction business magazines.

Get involved in playing golf. The trust and friendships built on the golf course are likely more effective in building business directly or indirectly than many marketing campaigns.

Many years ago, one of my primary competitors in the restaurant construction niche accomplished a marketing grand coup by filming the construction of a Hardee's restaurant in record time. A stationary time-lapse camera with a view of the building site was set up to film 24 hours a day. On Day 1, the camera recorded the building location on the jobsite. On the eighth day, it recorded the opening of the front door of the completed restaurant building including all interior and exterior finishes and paved parking lot. The project was completed in seven days, from start to finish.

Well, almost. While nothing can diminish this achievement, the description is incomplete in that the seven-day period did not include certain prefilming work such as staging of all building materials and other preliminary work. Building authorities cooperated to ensure practically immediate inspections as required during the process. An extraordinary number of workers were onsite, and every subcontractor was primed and totally

Marketing | **45**

committed to becoming part of a record-setting project and schedule. Certainly, much planning and coordination were involved.

Although this example is not practical, by any measure the feat is certain to impress anyone with any knowledge of construction and lend value to the contractor who did it. Imagine the promotional value of setting up a vendor booth at future quick-serve franchisor conventions and running this film continuously for the attending franchise owners to see and marvel at.

As a *niche* general contractor, I hope this real-life example impresses on you that specialization must be considered a viable route to follow in your construction career. The contractor, who is based in Atlanta, has had annual revenue figures of more than $250 million. Chapter 16 describes the advantages inherent in niche contracting.

By combining these materials and strategies, you can create a marketing program that effectively reaches and engages prospective customers, showcasing your firm's expertise and capabilities in commercial construction.

6

Subcontractor Management

In the construction industry, subcontractors are the backbone of any project. Their specialized skills and dedication ensure that every detail is executed to perfection.

Kevin O'Leary, Entrepreneur and Investor

Among your responsibilities as a general contractor, none ranks higher than that of managing your subcontractors. In carrying out a construction project the general hires subcontractors to do parts of the work, but the ultimate responsibility for carrying it out remains squarely on the general contractor's shoulders.

The policies and procedures outlined in this chapter are essential components of contractors for enduring success. Implementing them thoughtfully and effectively will help ensure your success in the selection, management, and performance of subcontractors.

Effective Subcontract Documents

In the early days and months of my construction business, I developed a cadre of reliable subcontractors and got busy before I had created or adopted any standardization in subcontract forms. The stationery-store forms I used were anything but enforceable or protective of my company.

This was a gap in my preparedness to be a general contractor that would never have occurred had I built an infrastructure for my company before starting business. In any dispute over a substantial liability due to, say, a jobsite accident that resulted in property damage or serious bodily injury or death, generic contract documents would not have been very protective of my company, the subcontractor, the owner, or the general public. That set of circumstances is a dream come

---

Mastering the Construction Startup: A Business Infrastructure Guide, First Edition. Nick B. Ganaway.
© 2025 John Wiley & Sons, Inc. Published 2025 by John Wiley & Sons, Inc.

true for the injured parties' law firms who would eventually become involved, and I would have received painful, unwanted lessons in litigation and the courts system, disruption of my fledgling business, and back-breaking legal bills. This is one example of things that can put startup contractors out of business and likely in debt and demonstrates the critical need for a well-drawn contractual agreement for all parties to the project. This is a component of your business infrastructure.

Before any of those consequences became a reality, I hired an attorney in a major construction law firm with whom I established a professional relationship that beneficially lasted for more than 20 years. A key benefit of that alliance over time was my coming to understand both legally sound and legally risky construction practices, including executive, administrative, and jobsite aspects. For larger or more involved projects, my company adopted industry-standard agreement forms mentioned earlier that may be tailored to the subcontractor and the project. For minor work, we used simpler but still effective in-house forms. In either case, the defining particulars of a given project must be incorporated into the form.

Having proper contracts in place between the subcontractor and contractor is no guarantee of easy resolution when a problem arises, but a well-thought-out agreement imposes methods, procedures, controls, and fixes on each party designed to reduce opportunities for things to go wrong; specifies insurance for a level of protection for the parties and others; and prescribes how default, disputes, claims, and lawsuits are to be processed. A well-written subcontract agreement can reduce dollar cost and time spent by the parties and their lawyers. The following chapters also offer effective practices and policies intended to minimize contractual disputes.

Even though contracts may vary in length, terms, and complexity, they must meet certain legal requirements that are described in Chapter 9.

Unless you are a full-service contractor performing all of your offered services with in-house employees, subcontractors will be a necessary resource requiring management and controls. Here are some of the steps toward better contractor–subcontractor relationships.

Trust

The contract signing or kickoff meeting between contractor and subcontractor is the time to establish ground rules for effective communication and contract compliance. A good start-off with a subcontractor is a firm handshake reinforcing verbal commitments, and a meeting of the eyes if proximity allows. You may view this step overkill, but despite being informal, I assure you this personal ceremony will be better remembered by the subcontractor than the commitments he agreed to in the contract document he just signed. It adds life, human touch, and personal pledge to the written agreement. This kickoff is most effective when it is you who conducts the meeting but if not, it should be one of your managers of authority the subcontractor will report to.

Adding the human interaction to the mix can be the difference in an effective resolution to a future issue or a dispute that leaves scars – both financial and relational – often to the detriment of the project. The arrogant *My way or the highway* style of management is the epitaph on many a contractor's bankruptcy monument. That does not at all mean you should not flex your contractor muscles, but effective management requires information, objectivity, emotional maturity, and intelligence.

Develop a fair and enforceable subcontract agreement and make it known to your project management and field employees that you expect them to strongly protect your company's interest, but at the same time try to maintain good working relationships with your subcontractors, vendors, and others.

In my experience, subcontractor problems most often arise from poor communication and understanding of the scope of work between the parties during the bidding process, change orders, scheduling, coordination with other subcontractors, and contract compliance. Dealing fairly and in good faith and expecting the same from the other party makes it more likely that potential problems are agreeably settled that might otherwise end up in court, which is almost certain to have a negative impact on the project and all parties.

Trust and cooperation between your company and your subcontractors will be enhanced by payment to the subcontractor in accordance with the subcontract, scheduling regular meetings with the subcontractor where small issues are recognized and addressed before they become bigger issues, and representing the subcontractor fairly to the owner.

Prequalification of Subcontractors

Qualify subcontractors to meet the unique requirements of your job before asking them to bid. Have they worked successfully on a project of similar dollar cost? Similar complexity? Similar manpower requirements? Similar pace?

50 | 6 *Subcontractor Management*

Substantially in the same geographic area? Have they worked on projects that require specialty materials and methods that may be required on your project? Do they have the necessary tools and equipment? Financial resources? How about their performance record with other general contractors and with their banks? Do they have a reputation for cooperation and meeting schedules? You need to know these things.

There may not be a perfect subcontractor fit for a given job. Your success relies on sorting out those whose deficiencies, once identified, are manageable.

The electrical contractor on a hospital project is a world apart from one for an office building in the scope of capabilities, number of employees, financial status, culture, and others. And no subcontractor is an island. Their work throughout the course of a project is choreographed with the other subcontractors. This is especially the case when working in limited space on smaller projects.

Failure to properly qualify a subcontractor with respect to type and scope of the project at hand may lead to delays in finalizing a bid, slow job progress, unqualified personnel on the job, inadequate number of available employees, poor workmanship, subcontractor money problems, project owner complaints, penalties against the contractor, and more.

You can improve your chances of selecting the right subcontractors for your projects by thoroughly describing the scope and duration of the project and soliciting multiple bids, which may single out bidders who are not right for the project.

Subcontractor Proposal and Scope of Work

The subcontract must clearly define the scope of work, which is a purpose of a well-drawn set of construction drawings and specifications. If a bidder on the project is aware of any significant omissions or errors in the documents, they are obligated to make it known to the owner or designers prior to the bid date so that the bid documents can be revised and uniform among all bidders. A subcontractor bidder who wins your job will not likely be able to claim a change order if it can be shown that they knew or should have known of the discrepancy prior to submitting their bid and didn't raise a flag.

Resolve any discrepancies between the subcontractor's proposal and the scope of work prior to contract signing, such as acceptable exceptions they made. Terms of the modification can then be incorporated into the subcontract document. For

Effective Subcontract Documents | **51**

expediency, on occasion, I have instead made the subcontractor's proposal part of the subcontract as an attachment, but this risks leaving potentially unrecognized differences to be confronted during the course of the project, possibly causing delay, or worse.

Include a performance timeline in the subcontract and formally notify the subcontractor of updates as conditions change. This can be done most effectively if you are using task dependency software.

Notice to Proceed Precaution

There will be special circumstances when the project owner needs work on the project to begin before all of the necessary preconditions are met, including contract signing and official notice to proceed. Every general contractor has faced this situation and leaned on to start work prematurely. This imposition on the general contractor could arise from third-party pressure on the owner, a last-minute glitch in a party's loan approval, a problem with the subcontractor's proof of insurance, last-minute delay in the issuance of some permit, and on and on.

You will be taking self-imposed potentially consequential risk if you start a project before an official notice to proceed is issued by the owner: Something could postpone the start of the project after work has been done, property damage is incurred, an employee becomes injured or killed, or any number of reasons. It is hard to imagine all the serious problems this could cause. Don't take this risk.

Official Notice

Separate from notice to proceed, contract agreements state exactly how notice of breach of or failure to comply with a contractual requirement is to be transmitted between the parties, i.e. names, addresses, time frame, etc. Even seemingly minor discrepancies can become an issue in case of a dispute hinging on whether notice was properly given or received.

It is wise to establish fail-safe procedures to ensure that official notices sent to you from the owner, or any other party, are time-stamped and delivered to your or your designated employee's attention immediately upon receipt. Expediency is necessary. Failure to respond to an official notice you receive from a party by the date specified may result in serious consequences.

Inspection of the Subcontractor Work

You or your project managers are responsible for routinely inspecting your subcontractors' quality of workmanship, especially at a time when their work is about to be covered up. In the event of deficiencies, document them with descriptions, photos, and related comments from anyone having direct knowledge and give timely notice to the subcontractor. Otherwise, any failure or liability that results from the subcontractors' faulty workmanship is likely to directly or indirectly fall into your lap.

Topographical Site Survey

For projects that include site work, the general contractor must be provided a current topographical survey (topo) for preparation of the site work bid by the project owner. The general contractor should provide copies of the topo to all relevant parties. Following is a list of features a topo survey may include, but all may not be necessary or applicable to a given project.

- Property boundaries: Clearly marked boundaries of the property, including any adjacent properties and legal descriptions.
- Contour lines: Lines indicating elevation changes and the shape of the land's surface. This helps in understanding the slope and terrain.
- Land elevations: Spot elevations at various points on the property to show the height above a defined reference point (often sea level).
- Natural features: Details of natural elements such as trees, shrubs, water bodies (lakes, rivers, and ponds), and rock outcrops.
- Man-made features: Structures such as buildings, roads, driveways, fences, utility poles, and any other constructed elements.
- Utilities: Locations of underground and aboveground utilities, including water, gas, electricity, sewage, and telecommunications.
- Hydrological features: Information on drainage patterns, flood zones, wetlands, and any other water-related features.
- Topographical details: Significant changes in terrain, such as hills, valleys, and embankments.
- Benchmarks: Reference points used for future surveying and construction activities.
- Existing infrastructure: Roads, pathways, parking areas, and any other infrastructure elements present on the property.

- Easements and rights-of-way: Legal restrictions on property use and areas designated for access by utilities or other entities.
- Soil information: General soil composition and conditions, which can be relevant for construction and landscaping.
- Legal descriptions: Official descriptions and documentation of the property's legal boundaries and characteristics.
- Zoning information: Details of zoning laws applicable to the property, including setbacks, height restrictions, and permissible uses.
- Geographical coordinates: Latitude and longitude coordinates for precise location referencing.
- Photographic documentation: Photographs of the property to provide a visual context of the surveyed area.

I once had a competitor who was halfway through construction of a free-standing restaurant job when he discovered that he was not building it on the designated property. I don't remember how it was resolved, but my people and I took a deep breath, realizing how easily this could happen by shortcutting procedures, such as failing to identify the property by use of a survey. Going forward we did not occupy a site before the boundary lines were staked out by a licensed surveyor.

Differing Conditions

A differing condition exists when planned features of a project conflict with what actually exists. This may be discovered during the bidding process or in a project under contract and in progress.

In preparing a bid, the bidders for various aspects of the site work should visit the site with the site plan identifying all proposed improvements and also topo drawings in hand and be certain that the work can be performed as required by the bid documents. An accurate site plan based on the topo will show the work to be done, but a bidder has the duty to notify the owner of any obvious significant site condition that differs from the topo survey or site plan. This could be a concrete monument partially buried and hardly noticeable, for example, or puddling or saturated ground, possibly indicating an underground spring or broken water or sewer line.

Upon discovery of a discrepancy between the bid documents and site conditions, the bidder should notify the project owner's designated

54 | 6 Subcontractor Management

representative. If the owner does not address the situation in revised bid documents prior to bidding, the bidder should seek guidance from the owner to not submit a bid that is disqualified for being noncompliant.

Whether or not indicated on the topo and site plan, the bidder should look for overhead utilities and cables that may need to be relocated in order to carry out the work. A utility pole might exist, say, where the site plan indicates a driveway onto the site. If it is a strategic utility pole whose relocation would have a domino effect on other parts of the utility system, it can cost tens or hundreds of thousands of dollars for the pole to be relocated. Here again, call this to the attention of the owner for guidance. Do not make the assumption in this case that the driveway can be moved over a few feet to avoid the conflict. Only the owner or architect can make that decision.

Another point here: if you are in the middle of a project and discover a design error, e.g. a drainage problem that you think can be easily corrected at practically no cost, do not do it without proper authorization. Let's say you do so and your fix inadvertently creates another more costly problem that is not apparent to you at the time. The owner or architect may try to hold you responsible for all associated costs on the grounds that your action precluded an appropriate fix designed by them, in which case the cost would be theirs, not yours. You may correctly guess that this caution is the result of the author's painful experience.

If a differing condition is discovered on a project under construction, it has the potential to become a very costly dispute. This further emphasizes the need that site investigation for bid purposes be conducted by you or a reliable employee experienced enough to recognize potentially significant features. This requirement applies also to subcontractors and others whose work could be affected by site conditions.

It is imperative to proceed methodically upon discovering differing conditions in the middle of a project. Immediately stop work related to the condition, notify the owner and others related to the condition, and take care to not do anything that can prevent determining the cause and the fix. No related work should proceed or be covered up. Take time-stamped photos and note the names of employees, subcontractors, and others who were involved or witnessed the situation. File this documentation. Differing conditions don't always become major disputes, but when they do, the party with the best documentation has better odds of prevailing.

Differing conditions are more common in renovation projects since the designers usually cannot know the conditions that may exist inside walls and other covered areas that prove to obstruct the planned work. Follow the precautions and procedures described earlier when they occur.

Independent Contractor Versus Payroll Employee

The classification of a worker as an employee or an independent contractor has significant implications for tax purposes and other reasons. The US Internal Revenue Service (IRS) uses several criteria to determine the correct classification.

Electrical, plumbing, and heating, ventilation, and air conditioning work is usually performed by licensed professionals or their company who contract with the general contractor to provide labor, materials, and supervision to perform a "turn key" job in accordance with the contract documents. They acquire their own permits and upon completion of the work arrange for their final building authority inspections, clean up, and move on. There is usually no question that such licensed firms qualify as independent contractors.

There are often circumstances when the general contractor needs services for which he prefers to hire people or businesses on a contract basis instead of the more involved course of hiring them as payroll employees when the need for them is limited and sporadic. This might be site cleanup, hauling, jobsite labor, minor construction, or renovation projects. It could include tradespeople such as carpenters for work where the general contractor provides supervision and materials.

However, the general contractor's treatment of workers or a company as independent contractors is likely to be carefully scrutinized by the IRS. Its regulations narrowly define who qualifies for independent contractor classification, but the rules are not clear-cut. The general contractor's inadvertent misapplication of the rules can lead to IRS penalties.

There are many advantages to the general contractor when hiring workers or companies as independent contractors instead of as employees. One such advantage is that the general pays no payroll taxes. Employers are not required to withhold or pay payroll taxes, including Social Security, Medicare, and unemployment taxes for independent contractors.

This means that independent contractors are responsible for setting aside part of their income to pay federal and state (if applicable) income taxes and self-employment taxes when due. This can be forgotten or delayed, possibly causing cash problems for the independent contractor at tax time.

6 Subcontractor Management

Employers are not obligated to provide benefits such as health insurance, retirement plans, paid time off, or workers' compensation insurance to independent contractors. Managing payroll, benefits, and other HR-related tasks for employees is time-consuming and costly.

Independent contractors can be hired for specific projects or tasks, allowing general contractors to scale their workforce up or down based on project needs without long-term commitments. Independent contractors often bring specialized skills or expertise that the general contractor does not have in-house. Unless there is a written agreement to the contrary, the general contractor can stop the cost on the day the independent contractor completes the work he was hired to do.

This generally suits independent contractors who often spread their services among multiple general contractors with whom they've established relationships. They like being their own boss and prefer independent status over being an employee. It is a win–win situation for both the general and independent contractors. The IRS is the only party opposed.

If the independent contractor does not have certain insurance coverage, he is usually covered by the general contractor's insurance, which results in the general's cost for insurance to increase, especially in the event of an insurance claim by the independent guy. The general contractor's insurance carrier makes this determination by way of its periodic audit of his records.

Independent contractors are also responsible for their own compliance with industry regulations and standards, reducing the burden on the general contractor.

Independent contractors further reduce the general's management requirements since the independent contractor usually provides their own supervision. Payments to independent contractors are typically tied to the completion of specific tasks or milestones tailored to motivate timely and performance standards.

These characteristics together reduce labor costs and increase operational flexibility, allowing general contractors to offer more competitive bids and improve their profitability.

IRS Regulations for Independent Contractors

As said, the IRS provides guidelines intended to determine whether a worker is an employee or an independent contractor, but the determinative rules often do not align with a contractor's specific circumstances. Despite this ambiguity, noncompliance with the rules can lead to problems for the general contractor.

The IRS determination for worker status is based on three main categories: behavioral control, financial control, and the relationship of the parties.

These metrics examine whether the company has the right to direct and control how the independent contractor performs the tasks. For example, does the company provide specific instructions on when, where, and how to do the work? The IRS also looks at whether the general contractor provides training to the worker or independent company.

Another IRS criterion is the general's degree of control over the financial aspects of the worker. An independent contractor usually has the opportunity to make a profit or lose money instead of wages. Other factors include whether the worker owns facilities, vehicles, and equipment he uses to perform the work for the general contractor and whether he makes his services available to the open market.

Workers' rights are a top reason for the IRS regulations. Employees are entitled to certain protections and benefits. The IRS claims that misclassification of employees as independent contractors may deprive or diminish workers of these rights and protections.

It is also concerned with fairness among businesses. Companies that classify employees as independent contractors avoid paying payroll taxes and employment benefits, potentially giving them a competitive advantage.

Misclassification can lead to penalties, back taxes, and interest. By monitoring and enforcing correct classification, the IRS seeks to deter misclassification and ensure that businesses adhere to tax laws.

Please see Appendix G for a sample independent contractor agreement.

Managing Change Orders

Change orders on construction projects are inevitable. It behooves the contractor to prepare for and establish policies and procedures for managing them toward the most favorable outcomes when they occur. Unmanaged, change orders often result in disputes, claims, schedule delays, quality issues, adverse relationships that may linger for the remainder of the project, and profitability.

There are any number of reasons in the course or a construction project that can trigger a change order, and the larger and more complex a project, the greater likelihood of change orders.

6 Subcontractor Management

By recognizing typical causes for change orders, you and your managers can be better prepared when potential change order circumstances occur. Follow a set of established procedures that will help determine cause, responsibility, cost, schedule changes, injuries, and liability. Here are a few common causes for change orders.

- Changes in design: Changes are common in prototype designs where design flaws or omissions are found only during the construction process. Design changes may also be required due to a change in scope. The owner of the project may decide midway to make modifications to better suit his needs. He realizes the main entrance lacks visual effect or a conference room is not large enough and directs the architect to make the changes and publish the necessary drawings, etc.
- Material shortages: Certain specified building materials may be discontinued, in short supply or modified in a manner unsuitable for the project. This may result in delay of the project.
- Unforeseen site conditions: Unforeseen site conditions can result from discovery of soil instability, subsurface objects or structures, underground utilities, or other causes that prevent carrying out the work as designed.
- Labor-related issues: Labor issues are more likely on union projects.
- Damages to the property: Weather conditions, mishaps, accidents, vandalism, and other causes can result in change orders.
- Code compliance: It is not uncommon for a deviation from applicable codes to be overlooked in the project design documents or during the authorities' inspection and approval process.

Effective Change Order Procedures

You can lower the likelihood of disputes by establishing standard procedures when the need for a change order arises. In most cases, you should stop work and temporarily freeze in time all related processes. Your jobsite and field employees should describe the situation in their phone or notebook for ready reference, including names, phone numbers, and workers involved. Documentation recorded at the time of incident is powerful in a dispute. However, no one in your company should accept fault or responsibility orally or in writing, which will be determined in due course.

Successful resolution of change orders avoids impact on project quality or schedule, leaves relationships between the parties intact, and may yield additional revenue to your company.

Still, neither the owner nor the general contractor wants change orders. They are time-consuming, often debatable, and in some cases delay the work. However, change orders are as common as the air on construction projects. Architectural drawings can have errors. Scopes of work can be erroneous or misinterpreted. Causes for a subcontractor's delay may be disputed.

Books have been written about change order disputes and their resolution by negotiation or through courts of law. In my experience, most change order disputes are resolved without major consequence.

7

Construction Disputes

If there's more than one way to do a job, and one of those ways will result in disaster, then somebody will do it that way.

Edward A. Murphy, Jr., American
Aerospace Engineer and Creator of
Murphy's Law

While construction disputes are often resolved without significant impact on the project, others involve high-dollar claims and lawsuits between project owner and general contractor, or general contractor and subcontractor, and spinoff claims stemming from the initial claim. Disputes may result in work delay or stoppage, largely proportional to the complexity of construction projects and the involvement of multiple parties.

Here are some of the causes for construction disputes.

Design Defects

Design problems are more likely to come to light during construction than in the design or bidding process. They often present themselves only after subsequent work has been done, which may cover up the problem, complicate the fix, and jack up the cost.

That is when the finger pointing begins. The contractor blames the project owner's engineering and design professionals. The designers throw the blame on the contractor for not following the plans and specifications and for making things worse by continuing with work that covered up the claimed design flaw. In addition, the project owner wants the designers to be

---

Mastering the Construction Startup: A Business Infrastructure Guide, First Edition. Nick B. Ganaway.
© 2025 John Wiley & Sons, Inc. Published 2025 by John Wiley & Sons, Inc.

right so that the cost to correct falls on the general contractor. Assignment of blame, angry meetings, hot emails, and legal notices are sure to impact the project.

Design problems may occur due to miscalculation and missing information, often in the design of mechanical and electrical systems. There can be one or more causes for this. Blueprints are sometimes the product of different teams, making communication and coordination issues more likely. A design professional may overlook or misinterpret a code requirement or incorrectly size rooms or enclosures, and more.

Workmanship Defects

Examples of poor workmanship include subpar soil compaction, excess water in concrete pours, faulty installation of parts or materials, incorrect concrete reinforcement, or roof leaks. Workmanship issues also include uneven floors, electrical and mechanical problems, stains, drywall cracks, foundation cracks, and door and window misfits.

Negligence

Construction defects may occur due to negligence of the contractor or others to meet industry standards of care, skill, and safety in the performance of their contractual work. According to the Occupational Safety and Hazards Administration (OSHA), falls are the most prevalent cause of death in construction accidents. Lack of or improper scaffolding and accidents involving ladders contribute to this. Code requirements may be missed during construction, as well as in the design process.

I am familiar with an incident on a project (not my company's, fortunately) where an employee working on the roof of a building made contact with a makeshift ungrounded extension cord while the roof was wet following a rain. Tragically, the employee did not survive. You can imagine the multifaceted impact of this incident on various parties and the project.

Nonshored trench cave-ins often occur due to contractor expediency or negligence despite building codes that require shoring at some depth. One or more employees may be partially or wholly buried by the collapsing trench wall and often do not survive frantic rescue efforts.

Another actual example of a tragic accident is one that occurred when a heavy object fell out of a moving dump truck onto the roadway pavement, took a high bounce, and struck a man riding in the back of a pickup truck approaching in the opposite lane. This accident would have never occurred if the driver of either of the trucks had been following proper safety precautions. The load of the dump truck should have been securely covered, and no person should have been allowed to ride in the open back of the pickup.

Changes in Scope

It has been reported that change in the scope of work was the primary cause for some 40% of construction industry claims and disputes in the United States during a recent 10-year period, second only to negligence.

This highlights the importance of accurately writing the scope of work in the original bid documents and tracking any changes during negotiations leading up to the final contract documents.

Changes in the Work

Changes that occur during construction due to circumstances including unforeseen site conditions, regulatory requirements, owner changes, and changes or errors in design are ripe for disputes and must be carefully documented. Unforeseen conditions are common in remodel or renovation projects.

Contract Terms and Conditions

A common issue that arises in construction is contractual disputes. These can arise from poorly drafted contracts, ambiguous terms, or disputes among parties over the intention of an item in the scope of work. I have my construction lawyer review contract documents before I sign them unless it is a form I am familiar with. This gives me peace of mind and less chance for unwanted surprises. I have learned from my lawyer that in a legal dispute, the party who drafted the agreement is held more closely to its terms.

Contractual agreements should define what constitutes default or breach of contract by a party and also specify the rights and duties of the respective parties in such an event.

Payment

Cash flow is vital for any construction project. Payment disputes arise from late payment, nonpayment, or disagreement over the value of completed work at the time of billing. These issues severely impact a project's timeline and the contractor's financial viability.

Methods of Dispute Resolution

Contentious disputes may be settled in several different ways.

Negotiation

Direct negotiation between the parties is almost always the fastest and least expensive way to settle a dispute, but for various reasons, it may not be successful or even attempted. The parties may have created a history of accusations and counter-accusations, the owner withheld payment, or the contractor stopped work. For valid cause or not, any one of these events is very likely to create anger and emotions that make rational debate hard to maintain.

Negotiation becomes less feasible in disputes involving multiple parties, complex issues, or multiple independent issues.

Mediation

In mediation, a neutral third party assists the parties in reaching a voluntary, mutually acceptable resolution. Mediation can be used to resolve any issues important to the parties, not just an underlying legal dispute.

Mediation has several advantages. The parties have the opportunity to address their particular needs and concerns and explore alternative solutions. These options are probably less available in a legal case.

Contrary to cases that end up in court, mediation proceedings are private. This appeals to parties who might have to publicly disclose sensitive financial or other information. The element of time also favors mediation. Court cases can drag on for a long time, complicating the project's normal course. Also, the parties are probably more likely to emerge from mediation without damaging their working relationship. Mediation can be binding if the parties enter a legally binding mediation agreement or if it is court-ordered.

Contracts often include notice and cure provisions, meaning that the claimant must give formal notice to the other party and the opportunity to cure a contractual breach by a certain time or date before initiating another formal dispute resolution method, including mediation.

Mediation is not always an option. Contracts can explicitly exclude mediation altogether. In others, parties may agree to forgo mediation and proceed directly to arbitration or litigation.

Contracts may also specify a particular jurisdiction where disputes must be resolved. You usually want to specify your home state if possible, giving you the advantage of using your own trusted attorneys. This also precludes you from having to pay your familiar lawyers for time and travel to another jurisdiction or, if not that, hiring lawyers in that jurisdiction whom you don't know and who don't know you.

Arbitration

Construction arbitration is a dispute resolution process in which an impartial third party, the arbitrator, makes a binding decision in a dispute. This method is often chosen to provide a quicker and more efficient resolution than may be possible in litigation. Contrary to court litigation, arbitration lets the parties establish many aspects of the process, such as choosing the rules for the procedure and protecting trade secrets and other sensitive business information, and is less formal. Among the advantages is the relative comfort of meeting in the privacy of a closed and conveniently located conference room in contrast to a public courthouse where security, added stress, and logistics come into play.

In simplest terms, the parties to arbitration agree on the selection of the arbitrator or panel of arbitrators, both parties present their case, and the arbitrator deliberates and issues a decision, known as the award. It is easy to imagine that a party's superior knowledge, preparation, documentation, and presentation could be dominant in the arbitrator's decision.

There can be more than one arbitrator in a case.

The speed of arbitration relative to litigation allows the contractor to get back to normal business sooner and without the dispute continuing to weigh on his shoulders, as well as on operations. But while arbitration is often faster and more efficient compared to litigation, particularly complex cases or those requiring multiple arbitrators and experts may be no less expensive. It is important

66 | 7 Construction Disputes

to select an arbitrator with knowledge specific to the industry or subindustry involved, which can lend credibility to the award.

If throughout the bidding and construction process, the parties strive to follow good policies and practices and communicate effectively, simmering disputes may be avoided or settled amicably. But if disputes persist, you will be prepared to make your case if you have maintained contemporaneous reports, photos, names of people involved, and other pertinent information.

8

Project Delay

Time and money are largely interchangeable terms.
Winston Churchill, former Prime Minister of Britain

Delay claims arise when unforeseen circumstances push a project past the initial agreed-upon completion or interim deadline. Construction delays are part of life for the general contractor and are costly in different ways to some or all stakeholders. This chapter looks at the usual causes for delay and describes steps the general contractor must proactively take to prepare for the inevitable delay.

Causes for Delay

- Material shortages: Delays in the availability of essential materials.
- Shipping delays: Problems in transportation logistics, leading to late delivery of materials.
- Price fluctuations: Sudden increases in material costs can cause delays while funding is reassessed.
- Labor shortages: Insufficient skilled labor to meet project demands.
- Labor strikes: Work stoppages due to labor disputes.
- Health and safety incidents: Accidents or health issues that reduce workforce availability.
- Owner-requested changes: Modifications requested by the project owner after the project has begun.

Mastering the Construction Startup: A Business Infrastructure Guide, First Edition. Nick B. Ganaway.
© 2025 John Wiley & Sons, Inc. Published 2025 by John Wiley & Sons, Inc.

8 Project Delay

- Design errors: Mistakes or omissions in the initial design that need correction.
- Approval delays: Sluggish approval processes for design changes.
- Adverse weather: Inclement weather such as rain, snow, or extreme temperatures that can delay construction activities or even damage the work in progress or completed including erosion.
- Seasonal constraints: Delays related to working within specific seasonal windows.
- Permit delays: Bureaucratic processing of necessary construction permits. This is more common in some jurisdictions than in others.
- Regulatory changes: New laws or regulations introduced mid-project requiring compliance adjustments.
- Inspections: Delays due to scheduling and completing mandatory inspections.
- Delay of funding: Problems in securing financing or disbursement of funds.
- Cost overruns: Budget issues requiring project rescheduling and financial restructuring.
- Unexpected site conditions: Discovery of unforeseen site conditions, such as compaction problems, buried debris or hidden utility components.
- Equipment failures: Breakdowns or unavailability of critical construction equipment.
- Complex engineering challenges: Difficulties in executing complex design elements.
- Poor planning: Inadequate initial project planning and scheduling.
- Communication breakdowns: Lack of effective communication among stakeholders.
- Coordination issues: Problems in coordinating multiple subcontractors or phases of the project.
- Scope changes: Disagreements over the project scope or contract terms.
- Payment disputes: Conflicts over payment terms and schedules.
- Environmental regulations: Compliance with environmental protection laws causing delays.
- Site contamination: Delays due to the need for environmental remediation.
- IT system failures: Problems with project management software or IT systems.
- Data loss: Loss of critical project data impacting schedules.

Be Prepared

Not all of the causes for construction delays can be prevented, but here are moves you can make proactively or upon occurrence that may prevent some delays and reduce the impact of others.

- Establish relationships with multiple vendors to reduce dependency on a single source.
- Develop strategic partnerships with key manufacturers to secure priority access to materials.
- Form joint ventures or alliances with other contractors to share resources and purchasing power.
- Be alert to early indicators of coming shortages of materials, which may allow you to stock up on those you anticipate that you will need.
- Implement effective inventory management practices to monitor stock levels and forecast future needs.
- Source alternative materials that meet project specifications and regulatory requirements.
- Work with architects and engineers to design flexibility into projects, allowing for substitutions without compromising quality or performance.
- Adjust project timelines to focus on tasks that do not require scarce materials.
- Implement phased construction to complete parts of the project that do not rely on the shortage-affected materials.
- Utilize project management software.
- Stay informed (e.g. trade and industry reports, publications, and newsletters) about market conditions, geopolitical events, and economic trends that could impact material supply.
- Include escalation clauses in related agreements that address price increases for materials due to shortages.
- Conduct risk assessments for each project to identify potential vulnerabilities related to material shortages.
- Develop a comprehensive risk management plan that includes strategies for mitigating material supply issues.
- Allocate contingency funds to cover potential cost increases due to material shortages.

8 Project Delay

- Secure financing options that provide flexibility in case of unexpected expenses.
- Maintain open and transparent communication with clients about the impact of supply shortages on project timelines and costs. Avoid sudden or unexpected notice to project owners of changes.
- Negotiate with clients for adjustments in project scope, schedule, or budget to accommodate the supply issues.
- Use advanced supply chain management software to gain better visibility and control over the supply chain.
- Learn and implement appropriate applications for various functions that can streamline processes.

As a startup general contractor, you are unlikely to encounter many of the causes for the delay described earlier. But as your projects grow larger, longer, and more complex, more and more of them will crop up and you will be better prepared to deal with them.

9

Contracts and Agreements

A contract isn't worth the paper it's written on unless it is backed by a strong sense of integrity and commitment.

Louis Nizer, Noted Trial Lawyer

As we go about living our lives, it is practically a daily occurrence that we enter into a contract with another person or entity. Virtually any purchase we make involves a contract. If we buy a car or truck, we enter into a contract with the manufacturer, the dealer, the insurance company, the tire maker, and the lender. A bank loan requires your signature on paper or electronic documents – a contract. Commercial websites often have a blurb saying that by clicking on a certain link you are *agreeing* to such and such. A contract is also called an agreement.

What Is a Contract?

In general terms, a contract is an agreement between two or more parties where each party assumes a legal obligation that is enforceable by law. Lawyers say that to be enforceable, a contract must comply with the following:

- The purpose and execution of the contract must be legal.
- A party's agreement to do or not do something in exchange for consideration.
- Acceptance by one party of the terms of an offer made by another party.
- The parties' ability to understand and carry out the terms of a contract.

The purpose of a contract is for the parties to agree on one or more conditions and obligations that are designed to result in a mutually desired product, service, or other outcome.

Mastering the Construction Startup: A Business Infrastructure Guide, First Edition. Nick B. Ganaway.
© 2025 John Wiley & Sons, Inc. Published 2025 by John Wiley & Sons, Inc.

9 Contracts and Agreements

The Associated General Contractors of America (AGC), the Associated Builders & Contractors of America (ABC), and the American Institute of Architects (AIA) all have semi-standard forms designed for use in practically every construction set of circumstances. Such forms simplify contract preparation by including stock language to cover a broad set of circumstances. It is left to you or your attorney to revise or add provisions that conform the form to your particular project.

A project owner's lawyers may elect to create their own contract form, possibly a version of which is used for the owner's other projects. Some of the terms of an owner-drafted contract will almost certainly tilt in favor of the owner in a close call. Be sure to have your construction lawyer review an owner-drafted agreement and negotiate for necessary revisions.

Either party may draw up the contract, but I prefer to have my attorney prepare the document. Any given provision of a contract may be interpreted to have a different meaning by two different readers. In the event of a conflict, be assured the lawyers on both sides will scrutinize every word for the slightest ambiguity.

Don't be tempted to sign an unfamiliar contract form without having your lawyer look at it. Just one legal dispute could cost you more than all the fees you would have paid your lawyer in your entire business career.

In the rush of things, you will at times be tempted to sign a contract without fully understanding all of its terms, and in most cases, the project will proceed to completion without a hitch. But if it is a project in which disputes arise, you want to be sure you are prepared. What seemed like a neutral provision when you signed the agreement may turn out to be harmful.

The Value of a Written Agreement

Any two or more honorable and well-meaning people entering into an agreement might orally agree on any combination of conditions, look each other in the eye, and shake hands to seal the deal.

So what could go wrong? A problem with such agreements arises when, for example, one of the parties remembers an important detail differently than the other party remembers it, and maybe both are being honest. Such disputes are often settled by a 50–50 compromise after much time and money have been spent by the parties arguing about it. This satisfies neither party and can be grossly one-sided.

When you contract with another party for products or services, your lawyer should insist that the other party indemnify you, defend you, and hold you free of harmful consequences from any claims stemming from the party's performance of the contract. This works both ways. The owner you are building for will require the same of your company. When problems arise, each party's attorneys will look for every possible reason to place the blame on the other side. You should have the appropriate insurance and corporate shield in place to protect you.

Types of Construction Contracts

The firm ConsensusDocs publishes a catalog of more than 100 documents that cover all or most contract document needs. According to the organization, its forms are the only standard contracts developed by a coalition of 40+ leading associations in the design and construction industry. ConsensusDocs contract forms intend to incorporate best practices and fairly allocate risk.

According to ConsensusDocs.org itself, the outfit "was created and developed in 2007 under the leadership of numerous organizations in the architecture, engineering, and construction (A/E/C) industry through a coalition effort with one goal: to transform construction by developing contracts and documents that protect the best interests of the project." Several types of construction contracts can be purchased from ConsensusDocs.

Lump-sum Contract

A lump-sum contract, also known as a fixed-price contract, is the simplest type of agreement between contractor and client. In this contract, the contractor agrees to provide specific services for a set price. This means the contractor takes on all the risk, so they might charge more to cover any unexpected problems. The contractor is responsible for completing the job properly and using their own methods and resources to do so.

This type of contract is usually created by estimating the costs of labor and materials and then adding an amount for the contractor's overhead and profit. If the actual costs are higher than estimated, the contractor's profit decreases. If the costs are lower, the contractor makes more profit. However, the cost to the client remains the same.

A lump-sum contract works well when the project's scope and schedule are clearly defined, allowing the contractor to accurately estimate the costs.

Unit Price Contract

In a unit price contract, the work is divided into different parts, usually by trade. It is based on estimated quantities of items needed for the project, along with their unit prices. The final project cost depends on the actual quantities used. For example, paving might be charged per square foot. While unit price contracts are rarely used for a whole major construction project, they are often used for subcontractor agreements where the types of work are clearly identified but the quantities are not specified in the contract. They are also commonly used for maintenance and repair work.

Cost-plus Contract

A cost-plus contract is an agreement where the project owner agrees to pay the full cost of materials and labor plus an additional amount for the contractor's overhead and profit. This type of contract is preferred when the scope of work and the types of labor, materials, and equipment needed are uncertain. In this arrangement, the contractor's profit is fixed. If the actual costs are lower than estimated, the owner saves money. If the costs are higher, the owner pays the extra amount. The main advantage of a cost-plus contract is that it usually ensures the project is completed as planned, even if costs increase. The builder is less likely to cut corners or use cheaper materials since their profit is not at risk.

Incentive Contract

A form of the cost-plus contract is a cost-reimbursement contract, which provides for the initially negotiated fee to be adjusted later by a formula based on the relationship of total allowable costs to total target costs. This type of contract specifies a target cost, a target fee, minimum and maximum fees, and a fee adjustment formula. After project performance, the fee payable to the contractor is determined in accordance with the formula.

A guaranteed maximum price (also known as GMP, not-to-exceed price, NTE, or NTX) contract is also a cost-plus contract, but the contractor is compensated for actual costs incurred plus a fixed fee subject to a ceiling price. The contractor is responsible for cost overruns, unless the GMP has been increased through formal change order (only as a result of additional scope from the client, not price overruns, errors, or omissions). Savings resulting from cost underruns are returned to the owner. This is different from a lump-sum contract where cost savings are typically retained by the contractor and essentially become additional profits. Sometimes, savings are shared between the owner and the contractor as an incentive to keep costs down.

Design–Build Contract

A design–build contract is appropriate when the project delivery method is design and build. With a design–build contract, the owner awards the entire project to a single company. It is typically awarded to a contractor, though architects or engineers may be awarded one in some specialized cases. Once the contract is signed, the contractor is responsible for all design and construction work required to complete the project.

When this type of contract is awarded to a contractor, they must hire all architects and engineers required to complete the design work. The owner is still given the right to approve or reject design options but is no longer responsible for coordinating or managing the design team. Once the owner approves the design, the same contractor then oversees the construction process, hiring subcontractors as needed. Most of these contracts are awarded through negotiation rather than through a bid process.

Traditional contracts are awarded using a design–bid–build system, where the project owner starts by hiring an architect. Once the architect has finished the design phase, the project is put out for bid to general contracting companies. The contractor with the winning bid is awarded the project and is responsible for completing the job according to the plans created by the architect.

Getting Paid

No general or subcontractor could honestly claim he's never had a concern about getting paid on one or more of the projects he has built. And for good reason: In the course of a months- or years-long project, there are opportunities for things to go wrong and when they do, the project owner may try to withhold payment until the problem is resolved.

Therefore, your lawyer should insist on contract language that aims to address nonpayment by the owner: The project owner must not be contractually entitled to withhold payment for work done prior to the disruption for which you are not responsible. But the owner will often do so anyway and argue about it later.

The AGC, ABC, and AIA contract forms *attempt* to cover such contingencies, but your lawyer should confirm that the nuances of your particular project are adequately addressed. But dotting every *i* and crossing every *t* does not guarantee that the owner can or will comply. Disputes often wind up in court regardless of the best intentions of all the parties involved, or how well a contract is written.

As example of an extreme case of delayed payment by the owner, my construction firm years ago contracted to build a Burger King (BK) restaurant on a site that had been occupied by a gasoline station. There were three parties involved in the project: Burger King corporate, who coordinated the bidding process and the subsequent construction project; BK's franchisee, who was the owner of the project and responsible for payment; and my company the general contractor.

BK took the role of manager for the project for its franchisee, but the construction contract was between the BK franchisee and my company.

An engineering report of the subsurface conditions (a *soils report*) was obtained and paid for by the franchisee, which my company was entitled to rely on as part of the contract documents. The report reported no buried objects that could interfere with the project as planned.

BK obtained all necessary permits and approvals, the contract was signed by all parties, and BK issued my company formal notice to start the project.

A month into the project we discovered discarded tires and other debris embedded in the soil, which, of course, risked future settlement of the improvements. In compliance with the contract documents, we immediately notified BK and the franchisee of the condition, and BK issued an official stop-work notice.

Getting Paid | **77**

We billed the owner for work to date in accordance with the contract terms. The franchisee refused to pay the invoice, saying he had done everything he was required to do by his franchise agreement, including paying for the soils report prepared by a licensed engineering firm. That is a strong argument in his favor, but we had performed the work in accordance with the contract documents, including the soils report provided by him, and were entitled to payment.

We stood by during the several months it took BK, the franchisee and other parties to work this out but placed a lien on the property for protection. A lien is a legal claim against the asset if its owner defaults. Lien laws vary from state to state and usually require a lawyer to file a lien properly.

When Burger King eventually issued notice for us to proceed with the project I agreed to do so only upon (1) receipt of the amount due for work we had completed before the stoppage plus interest, (2) Burger King and its franchisee's agreement to the revised contract amount based on the changed conditions, and (3) receipt of an irrevocable letter of credit issued by a bank of my choosing for the projected cost to complete the contract plus 20% to allow for any subsequent disputes, and payable to my company simply upon demand.

The necessary bank credit was established by the franchisee, who authorized the bank to pay a portion or all of the stated amount in accordance with the Letter of Credit language prepared by my lawyer. The bank required the franchisee to maintain account balances sufficient to cover the guaranteed amount.

The Letter of Credit provided that in the event of any delay of payment by the franchisee, I could collect the amount due from the bank without further authorization by the franchisee.

The franchisee reluctantly agreed to these one-sided terms knowing he would have paid a steep premium to get another general contractor to take the project over in this condition. I demanded the strident terms because I had lost all trust in the franchisee. Upon completion of the project, we collected the full amount and paid the unused balance back to the franchisee.

While the above is an unusual case it is not unrealistic, and there are provisions you can include in the contract that are likely to improve your chances of getting paid. Specify the time, event, state of progress or other metric that will trigger interim payment or payment in full. This applies whether you are the buyer or the seller of the product or service.

State clearly in the agreement the condition(s) to be met by each party prior to any partial or final payment. This might include receipt of a certificate of occupancy issued by the permitting authority, final inspection, completion of punch lists, or any condition you, as payor, deem effective under the circumstances. Specify the terms of any warranty or state that no warranty exists.

The payment due at the established milestone is often based on your cost to date versus the total contract amount. You can create other methods as needed, tailored to the particulars of your project. The project owner may perform his own calculation of payment due based on the agreed trigger to protect himself from overpayment.

The agreement should establish where invoices or payments are to be sent, such as a certain department, and, if possible, the name of a designated person to submit them to, and by what method – email, electronically, fax, United Parcel Service, etc. (do not depend on the US Postal Service) This procedure creates the opportunity for direct communication and an effective working relationship between the employees in the payables and receivables departments of both parties.

Following submittal of your invoice to Jones Company, you or your bookkeeper or accounting department should reach out to the designated Jones employee to confirm their receipt. Without these steps, you have no way of knowing if it arrived there, there's a processing delay, the owner is having cash flow problems, or your invoice is buried under a pile on someone's desk.

Include in the contract the method by which any change in completion date or contract price is to be determined if either is affected by internal or external circumstances.

The agreement must define what constitutes contract default by a party and what that party may or must do, and by when, to "cure" (relieve) the default.

Specify the state whose laws will govern in case of legal dispute and, as said elsewhere in this book, specify the state of your headquarters.

Your decision whether to include any or all of these provisions depends on several aspects of the transaction such as project dollar value, complexity, and duration. It is not unusual for one or more of them to come into play between you and another party, and when one does, you will be grateful that your written agreement thoroughly addresses it. In the hypothetically perfect contract, virtually no circumstances could arise that are not covered by the terms of the agreement. Perfection is not reality, but it is worth your time to cover as many potential areas of dispute as possible.

Of course, if the owner runs out of money you become best friends with your lawyer.

Example Lawsuit

This fictitious lawsuit exemplifies the importance of clear contract terms, thorough inspections, proper supervision, and adherence to contract requirements in construction projects.

Legal Brief

Case Name: Smith v. Zenith Construction, Inc. and Delta Foundations
Court: Superior Court of Fulton County, Georgia
Case Number: 2027-CV-12345
Filed: April 1, 2027
Plaintiff: John Smith
Defendant: Zenith Construction, Inc.
Third-Party Defendant: Delta Foundations

Narrative

Introduction

John Smith, a prominent restaurateur in Atlanta, contracted Zenith Construction, Inc. to construct a new restaurant. Zenith Construction subcontracted Delta Foundations to prepare the ground and pour the concrete building slab, including the installation of utility lines beneath the slab. This case arises from alleged failures in Delta Foundations' performance and Zenith's oversight, resulting in significant operational disruptions and financial losses for Mr. Smith.

Factual Background

In early 2026, John Smith engaged Zenith Construction, Inc. to build a state-of-the-art restaurant. The project required typical preparatory groundwork and the pouring of a concrete building slab under which essential utility lines – water, sewage, and electrical conduits – were to be installed. Zenith Construction subcontracted this work to Delta Foundations, a company specializing in such foundation work.

According to the contract documents, Delta Foundations was required to perform its work strictly in accordance with the specifications provided. Zenith claimed that Delta was contractually obligated to seek and receive Zenith's prior approval before pouring the concrete slab. Delta encountered unexpected soil conditions during the installation process but proceeded without notifying Zenith Construction, believing the slab preparation was sufficient to meet the specifications.

Upon completion of Delta's work, the project continued without immediate problems, and the restaurant opened on schedule. However, shortly after opening, Mr. Smith began experiencing severe operational issues. Water began pooling in the kitchen, electrical outages became frequent, and a persistent sewage odor plagued the dining area.

An investigation revealed the root cause of these problems was Delta's failure to address the unexpected soil conditions properly prior to pouring the slab. Delta blamed Zenith for failing to inspect and approve the slab pour.

Legal Proceedings

Faced with estimated repair or replacement costs, lost business, and reputational harm, John Smith filed a lawsuit against Zenith Construction, Inc., seeking to be made whole. Zenith Construction, in turn, filed a third-party complaint against Delta Foundations, alleging breach of contract and seeking indemnification for any damages awarded to Mr. Smith. Zenith asserted that Delta was obligated to seek and receive its prior approval before pouring the concrete slab, which Delta failed to do.

The case proceeded to trial, where Delta Foundations argued that while deviations were necessary due to unforeseen soil conditions, it believed they were acting within the scope of their expertise and responsibility. Zenith countered by asserting that Delta's failure to obtain its prior approval was a clear breach of its contractual obligations.

Court's Findings

The court examined the contract documents and found ambiguity regarding the requirement for Delta to seek Zenith's prior approval. Since Zenith had drawn up the subcontract, the ambiguity weighed against Zenith. The court found Zenith and Delta jointly and equally at fault for the issues that arose. Delta's failure to seek approval, its faulty work, and Zenith's unclear contractual requirements all contributed to the issue.

John Smith was awarded the full amount of costs and damages for the loss of the building he contracted for, lost business, and reputational harm, with Zenith and Delta each responsible for half of the total amount due to Smith.

Conclusion

This case underscores the critical importance of clear contractual provisions and the necessity for all parties to follow agreed-upon procedures. It highlights the need for explicit communication and approval processes in construction projects to prevent such disputes and ensure successful project outcomes.

Part 4

Resources

10

Lawyers

The job of a lawyer is to make sense of the confusion. To solve problems and resolve disputes. To be an advocate and an advisor. To help guide clients through the thicket of rules, regulations, and laws that abound.
Eliot Spitzer, American lawyer, politician, and
former governor of New York

The American perception of lawyers is probably not net positive, which may be because the lawyers who are most often in the news are those who sometimes represent egregious characters accused of committing horrific crimes. But we hear little about the lawyers who quietly and professionally go about the business of helping individuals, businesses, and organizations achieve their objectives and strategies within the guidelines of the laws and regulations imposed by federal, state, and local governments. I have relied on my lawyers extensively throughout my business career when faced with unfamiliar circumstances.

The US legal system is so broad and complex that many lawyers, much like doctors, specialize in the more complex areas of the law.

As a general rule, I recommend using professionals who specialize in areas in which the situation at hand falls. An attorney who specializes may or may not be smarter than a general practice lawyer but is likely more knowledgeable in a given area of the law based on their undergraduate or law school training, their field of practice, or both, which may better prepare them to see what's around the next corner before you get there. My long-time construction lawyer graduated from college with an engineering degree before becoming an attorney and joining the construction law firm I used throughout my years as a general contractor.

Following is a brief description of various types of lawyers.

Mastering the Construction Startup: A Business Infrastructure Guide, First Edition. Nick B. Ganaway.
© 2025 John Wiley & Sons, Inc. Published 2025 by John Wiley & Sons, Inc.

Construction Industry Lawyer

As a building contractor, subcontractor, or other participant in the construction industry, your needs for legal advice may often be simple enough to be answered in a quick call to your lawyer or it could be that a jobsite incident has occurred, causing bodily injury, death, or property damage.

In the second case, do not be tempted to call your general practice attorney. In such instances you need the services of a specialist in construction law, and you need them now. If you follow the advice in this book, your construction lawyer's phone number will be right there in your phone.

Many things can happen in the immediate aftermath of a serious incident, and getting your lawyer involved at the earliest stage can make a difference in the outcome. Your attorney surveying the situation on-site will know what to look for and be better prepared to handle the case. Even what initially appears to be a minor incident can morph into a major dispute or lawsuit.

When a serious jobsite accident occurs, especially if injuries are involved, TV and newspaper reporters often swarm the jobsite where they approach the first persons they see and ask questions. If your employee offers an opinion or even a description in the heat of the moment, it may appear on TV. In an ensuing lawsuit, such offhand comments might be used against your company.

In your attempt to prevent this from happening, it is good policy to periodically instruct or remind your employees to make no statements of any kind to the press or others if an incident occurs. Even you as owner of your firm might unwarily make a comment that is later used against you in legal proceedings. Regardless of the pressure on you, avoid making any comments before talking to your attorney.

A complex case often involves a large sum of money, job delay, payment issues, and contractual terms and interpretations. If it involves severe bodily injuries or even more serious claims, you certainly need a dedicated construction-law firm that has been around awhile, preferably one you have already formed a relationship with.

As contractors we hope it never happens but if the worst occurs you will almost certainly be better served by a regional or national law firm, which is more likely to have the financial ability and number of lawyers to hang in with your company for the duration. Even the best law firm with few attorneys can be drowned in legal paperwork, hearings, and depositions, which might even be the strategy of the opposing law firm. A major well-established law firm will have the layers of attorneys, trial experience, depth of experience, and the reputation it has built over years in the construction area. The lawyers of your opponent know this about your law firm and will understand there are no easy pickings here. National firms like Smith Currie Oles, in business since 1965, is one of many reputable construction law firms with offices located around the country.

General Practice or Business Lawyer

A general practice attorney will help you with the necessary local, state, and federal regulations in the process of forming your LLC or other type of entity, to prepare the necessary documents, and to guide you in most other routine business matters. Often, though, even these lawyers may have a preference for different areas of law, so it is advisable to inquire about his experience as it relates to your requirements in the course of choosing one to help you.

Circumstances may arise outside of construction for which you should consider a lawyer who specializes in a certain field of law. As in choosing a doctor who specializes in a condition you might encounter, you want a lawyer who knows their way around the circumstances at hand. Following are several legal specialties.

Employment Lawyer

An employment lawyer typically provides advice about legal issues arising from an employment contract or within an employment relationship, such as when terminating employment, which can be legally treacherous.

Workers Compensation Lawyer

In the case of an employee who has a work-related injury or illness, a worker's comp attorney is likely to become involved and may represent either the employee, employer, or the worker's comp insurance carrier. This complex area demands the expertise of a lawyer who specializes in workers compensation (WC) claims.

As a general contractor hiring subcontractors, for example, your WC insurance carrier will audit your books annually to determine whether you have employed subcontractors or others who did not have adequate WC insurance. If your WC policy covers your uninsured or underinsured subcontractors in addition to your employees, it will increase your premium to include the coverage provided to them under your WC policy that exceeds their own coverage. Your construction management software can help you manage your WC insurance and streamline WC audits. WC insurance is complex and requires careful attention to and management of claims made against it.

Real Estate Lawyer

A commercial real estate lawyer is hired to protect the rights of a party or parties in matters related to corporate ownership, leases, title insurance, and ensuring that the deal is legitimate. These lawyers prepare or review all of the paperwork involved in buying or selling property and may represent the lender, the buyer, or the seller throughout the process and at closing.

Mergers & Acquisitions Lawyer

If you now or in the future explore purchasing or merging with another firm, you will likely bring in a Mergers & Acquisitions lawyer to keep the playing field level. These lawyers help strategize, negotiate, assist with financing, draft legal documents, and execute transactions where two or more businesses combine into a single new firm or where one business purchases another.

Civil Litigation Lawyer

If you're suing someone or responding to a lawsuit against you or your business, you will need an attorney who specializes in civil litigation.

Tax Lawyer

Tax lawyers specialize in legal issues you may need help with if you have an issue regarding federal, state, or local tax laws.

Intellectual Property Lawyer

If you or your company owns intellectual property (IP), an IP attorney can advise you regarding protection of trade secrets, industrial design, patents, and other such property.

Personal Injury Lawyer

This type of attorney specializes in obtaining compensation in the form of damages for injuries caused by other parties or in defending against the same. In case of a personal injury that is claimed to have occurred on your jobsite, for example, you will become acquainted with a personal injury attorney or firm.

Immigration Lawyer

If you're dealing with immigration issues, such as visas, green cards, and citizenship, not unusual in the construction industry, you may want to consult with an immigration lawyer.

Estate Planning Lawyer

Estate planning lawyers specialize in wills and trusts. An estate attorney keeps current on the frequent and often far-reaching changes made by federal and state governments that affect how your estate will be inventoried, valued, taxed, and distributed after your death, and help with the probate process.

Deal Making

If your business now or later involves entering into financially significant deals, trust no one other than yourself and your lawyer and accountant. I have known intelligent people who negotiated business deals involving large sums of money and long duration and who trusted the other party and its lawyers to put the agreed-upon terms and conditions into the final contract document. This is unsound because there are nuanced meanings in almost any agreement that

90 | *10 Lawyers*

can depend on interpretation. It should be your policy to have your own attorney determine that the document says what you intend it to say before signing it. Do not make business deals using an attorney who is supposed to represent both parties. Get your own attorney involved.

While it is necessary to choose the kind of lawyer you need for your particular circumstances, it's equally important to choose one you are comfortable with at a personal level. In addition to being satisfied with his professional qualifications and experience common to your business, you want an attorney who will dig into and take a genuine interest in your situation.

You want to feel comfortable discussing it with him, as well as becoming confident, that he is not squeezing you into an overcommitted caseload.

Finally, if the prospective lawyer clears those hurdles, apply this acid test: Would you be at ease having a beer or sharing a leisurely meal with him or her?

11

Accounting and Taxes

A business without good accounting is like a car without a gas gauge – you might be able to drive for a while, but you're going to run out of fuel eventually.

Anonymous

If you're just getting started in a new construction business, you may plan to handle bookkeeping yourself to keep expenses down. And it is the nature of the startup owner or businessperson to want to be in charge of every detail, especially in the early stages.

But as your business grows and evolves and you are juggling more and more balls, your finances gradually become more challenging, adding complexity to the recordkeeping process. Mistakes and delayed entries are more likely and some of them, especially as they relate to cash flow, can cost you in many ways. But by paying attention to a few key warning signs, you can become immediately aware when you need help with your bookkeeping so that you can correct course.

The first warning sign you may notice is that your records are not being kept current, and if not addressed this can accelerate. To make sound and sometimes quick business decisions, you need up-to-date records that you can access on the fly and trust they are current. This includes bank balances, payroll records, billings, payables, and more. In addition to giving you confidence in your numbers, it will also help avoid problems when it is time to gather the records needed for filing your income taxes.

Decide early on how you will organize and maintain all these records. If you don't, they may end up being thrown into a drawer as they accumulate. Imagine facing this at the end of the year when you have to plow through all of them just to find an invoice or other document you need. Or even at the end of the

---

Mastering the Construction Startup: A Business Infrastructure Guide, First Edition. Nick B. Ganaway.
© 2025 John Wiley & Sons, Inc. Published 2025 by John Wiley & Sons, Inc.

11 Accounting and Taxes

month. Hiring a qualified bookkeeper as early as feasible will pay you back in multiples in terms of accuracy, timeliness, and access to your records, as well as more time you can spend managing your business. At this early stage, you may be able to accomplish this by hiring a bookkeeper for a few days a month.

The second red flag that you need help with bookkeeping can come at the end of the month. A critical part of your accounting and recordkeeping is prompt reconciliation of your bank statements, which is to confirm that the statements and your cash accounts line up with each other. When you are wearing several different hats, it is too easy to put this essential duty off until tomorrow and tomorrow. But timely maintenance will give you confidence in making financial decisions in the coming month. If there is a discrepancy, you could find yourself making decisions based on incorrect or incomplete information (I recount a significant personal example of this scenario in Chapter 18.). This should prompt you to bring in bookkeeping help sooner rather than later.

Another indicator that you need to bring in outside help with the books comes when you begin to expand or grow your business on the way to a new level. Keeping up with your finances, bank statements, invoices, payroll, etc., is becoming more complex – and harder for you to find the time to give it the priority attention it demands. You may be hiring new employees, seeking new working capital, or taking on a new customer or client. All of these require time and focus. All together, they make your bookkeeping process more complicated – and are more evidence that you need bookkeeping help. Continuing to go it alone may prove to be counterproductive. You could be triggering larger issues that grow fast if not given priority attention.

Of course, you have acquired construction industry accounting software that is widely available, but it doesn't operate itself, and it doesn't produce reliable records based on disorganized, incomplete, or inaccurate input data.

One of the first bridges to cross on the road to business success is to get the best person you can hire for each key position, and the employee who will have a lot of responsibility for handling your financial function is at or near the top of that list.

Tax Preparation

The size and complexity of your business will be a primary factor as you decide between doing your taxes in-house or using outside professionals.

According to a recent article in the *Wall Street Journal*, the average time all Americans spend preparing to file taxes each year is 13 hours. Business filers average 25 hours. I think the time spent will be greater for businesses as nuanced and complex as are many construction firms. Only while your construction firm is still relatively small and uncomplicated should you attempt to do your taxes yourself.

Some companies you work for may require annual, audited financial statements, which increases your accounting costs significantly. This may also apply when obtaining a contractor's license in some states or municipalities and when obtaining a performance and payment bond.

If you are accustomed to preparing your personal tax returns and your business income can be reported on your personal tax return, as is the case of a single-owner LLC, for example, tax preparation and filing your combined business and personal taxes may be a viable option. However, if your circumstances include complexities such as operating in multiple states or real estate transactions, you will probably need a professional tax preparer.

From the beginning, I have relied on my outside CPA firm to prepare my taxes. The single time that I or a business I owned was selected for an in-person Internal Revenue Service (IRS) audit was conducted in my CPA's offices, a small group that had established a long history of integrity and professionalism with its clients, as well as with the IRS. When I arrived for the audit, my CPAs had my records stacked neatly on a conference table and although we had been told it would be an in-depth audit, the auditor surprisingly spent less than an hour examining only a few of the records before declaring that no changes in my returns were warranted. I believe the apparent professionalism, preparation, and presentation by my CPAs were a major contributing factor to this welcome outcome.

My CPA did not want me present even for the introduction to the IRS agent, which is typical, because the accountant fears the client will inadvertently make an offhand comment that raises the agent's curiosity. But I wanted to meet this IRS person eye to eye, who to some extent held my well-being in his hands. But I didn't linger. I left the office shortly after the introductions.

Had the IRS auditor found something of concern, my tax-oriented CPA firm had the horsepower to defend my returns and the firm's work.

As important as any factor in selecting your CPA or other external accounting professional is your ease and comfort working with them. Decide whether they are a good listener and look for other clues to their personality and style

11 Accounting and Taxes

as they relate to your personality and other preferences. Visit their offices. How interested are they in taking you on as a client? How much experience do they have with construction accounting? Include the answers to these questions when making your decision.

For people like me who do not have an accounting background, it may require some research as well as initial input from your CPA for you to maximize the potential value of your financial statements to you. These documents are fundamental for tracking the financial performance of a business, making strategic decisions and ensuring financial health. The following are descriptions of the three primary components of a set of financial statements.

The Income Statement (Profit and Loss Statement)

The income statement provides a summary of the company's revenues, expenses, and profits over a specific period, which can be a quarter, six months, or full year. This statement is vital for assessing the company's operational efficiency and profitability. The income statement comprises the following.

Revenues: Total earnings from business activities, which in construction could include earnings from projects completed during the period.

Cost of goods sold (COGS): Direct costs attributable to its operations, i.e. the production of the projects built and other goods or services provided by the company, including labor, materials, and overhead directly involved in construction projects.

Gross profit: This is Revenue minus COGS. It reflects the profit a company makes after deducting the costs directly associated with its operations.

Operating expenses: Costs related to the operation of the business that are not directly tied to a specific construction project. This includes rent (or debt-servicing cost if you own the property), marketing, administrative expenses, and salaries of staff, which includes you, the owner. As a general contractor, you can look at this number as your cost to keep the doors open whether you have any projects in progress or not. It can keep you awake at night during dry spells, but it also identifies expenses that can be reduced or cut. Reducing staff is a fast way to lower expenses, but it may also reduce your capacity to generate profits when business comes back, and it has a painful human factor.

Operating income: Also known as Earnings Before Interest, Taxes, and Amortization (EBITA), it is calculated by deducting operating expenses from gross profit.

Net income: The final bottom line, calculated after deducting all expenses, including taxes and interest, from operating income. This figure shows the company's profit or loss for the period.

The Balance Sheet

The balance sheet is a snapshot of the company's financial condition at a specific point in time. It details the company's assets, liabilities, and owner's equity, showing what the company owns and owes. Here are the components of the balance sheet.

Assets: Resources owned by the company that have economic value. This includes current assets such as cash and inventory, and fixed assets such as land, buildings, and equipment.

Liabilities: The company's debts or obligations. These are categorized as either current liabilities (due within one year) or long-term liabilities.

Owner's equity: Also known as shareholders' equity, which is the company's total assets minus its liabilities. It includes funds contributed by owners and retained earnings.

Statement of Cash Flows

This statement tracks the flow of cash in and out of the business' operations, investments, and financing. It's necessary for understanding the company's liquidity and cash management practices. Here are the components of the statement of cash flows.

Operating activities: Cash flows from the core business activities. This includes cash received from customers and cash paid for expenses such as materials and payroll.

Investing activities: Cash flows associated with the purchase and sale of assets such as equipment and property, as well as investments not related to the core business operations.

Financing activities: Cash flows related to transactions with owners and creditors, such as receiving loans, repaying debt, issuing stock, and paying dividends.

Together, these three statements provide a comprehensive overview of a construction company's financial status. They allow interested parties to analyze the company's profitability, financial condition, liquidity, and cash management efficiency. They are also an essential part of informed management decisions.

Summary

When reading your financial statements or those provided to you by your customers and others, remember that the numbers represent that company's financial condition at a certain point in time, that shown as the date of the statements. The worn-out cliché, *past performance is not indicative of future results*, should always come to mind when you're analyzing financial statements – whether yours or those of another company.

The numbers and results indicated on financial statements are only as accurate as the numbers used in their preparation. Input inaccuracies can be the result of, for example, careless mistakes, delayed reporting of income or expenses, or intentional manipulation.

Banks and others who depend on your company's financial health may require *audited* financial statements by certified public accountants or firms (CPAs), whose job is to verify the accuracy of the numbers used in your financials. Audits signed off by CPA firms are universally relied upon by stakeholders and are expensive to produce. Consider yourself fortunate in that regard if you are not required to provide audited statements. Nevertheless, you should strive for complete and accurate financial reporting, which is imperative for good decision-making.

You must understand the source of every line item on your financial statements and their implications for your business. They can tell you what went right or wrong last year or last month and identify areas needing your attention. Another use for accurate financials is to identify your customers who are more – or less – profitable than others as a guide to where to employ your resources.

Financial and accounting software are essential tools to have from day one. There are many different software systems including those that are construction-industry-specific. Investigating these should be part of your early diligence.

Become familiar with your systems' abilities and limitations. Even if you are not the primary user, you will be more prepared to spot any errors and inconsistencies.

12

Hiring the Right People

Clients do not come first. Employees come first. If you take care of your employees, they will take care of the clients.

Richard Branson, British entrepreneur and
founder of the Virgin Group

Robert Iger, CEO of Disney, the movie and entertainment giant, might be the first to tell you that there's no way you can be certain you have chosen the right person for a given position. Disney in 2023 fired the man Iger had chosen as his successor, who had risen through the ranks at Disney for *26 years* before taking over from Iger as CEO. He lasted just two years before Disney fired him and brought Iger back in.

If an international corporation in the top tier of Fortune 500 rankings, likely having among the most sophisticated human resource systems in the corporate world, can't always successfully evaluate how even a person with a proven successful tenure will perform in a given role, how can small-business people like you and me expect to make consistently good hiring decisions?

The unfortunate reality is that any given prospective employee can check every box on your qualifications list and yet turn out to be wrong for the job. I have hired people after thorough vetting who didn't work out and conversely in my early years hired people on a gut decision who became my most valued.

Difficulty in hiring is no excuse for throwing up your hands in surrender, but is instead the reason that you have to make successful hiring a top priority and go all out to maximize hiring effectiveness. If you are striving for excellence in your company, as you should be, you will not likely make a more important decision in your company's future than in selecting the people who will fill

Mastering the Construction Startup: A Business Infrastructure Guide, First Edition. Nick B. Ganaway.
© 2025 John Wiley & Sons, Inc. Published 2025 by John Wiley & Sons, Inc.

98 | *12 Hiring the Right People*

the critical jobs. Like Disney, you won't always win, but you will beat the odds hands down.

And as the entrepreneur, business executive, and former Presidential candidate Ross Perot once said, "Eagles don't flock, you have to find them one at a time."

During a financial crisis that my company suffered, which I describe later in this book, the importance of hiring the best people for the job was brought home to me in vivid color and pain that kept me awake at night for more than two long years. The initial problem had several parents but the proverbial buck stops with me. To emphasize the theme of this chapter, it is fair to say that an effective accounting manager would have had safeguards in place that would have lessened if not precluded my company's financial crisis. And it was I who hired her.

So, Who Are the Right People?

The nation's economic conditions often determine whether prospective employees or businesses have the upper hand in determining compensation and benefits. Improving economies usually create businesses' need to hire, giving the job seeker the advantage.

At the time the earliest so-called Gen Z'ers (people born between ~1997 and 2012) were entering the workforce, national economic conditions created high demand for this generation and empowered them to a greater extent than earlier generations. Hiring later slowed during the COVID-19 pandemic, giving the greater bargaining position to the employer.

Many prevailing employee attitudes, values, social views, and expectations accompanied the Gen Z'ers into the American workforce. To varying degrees, significant adjustment will be required of many of us who are a generation or more older than the Gen Z'ers in how we adapt to, hire, and manage our workforce. The Gen Z group represented one-fourth of the US population in 2023, according to the *Washington Post* (Table 12.1).

Offering benefits and other features to help you attract and/or retain employees does not guarantee hiring success, but it will go a long way toward achieving your goal.

Attractive employment opportunities offer flexibility, technological integration, career development, meaningful work, and a balance between professional and

Table 12.1 Examples of strong company values.

Leadership	Trust	Accountability
Integrity	Innovation	Passion
Respect	Teamwork	Transparency
Quality	Constant improvement	Curiosity
Simplicity	Humility	Excellence
Perseverance	Diversity	Collaboration
Honesty	Craftsmanship	Service
Responsibility	Customer focused	Positivity
Courage	Community	Sustainability
Selflessness	Optimism	Environment
Ownership	Safety	Hospitality
Ethics	Celebrate success	Expertise
Communication	Learning	Professionalism

personal life. Gearing up your hiring processes and evolving your company culture to meet these expectations will improve your ability to attract and retain talent from the newer workforce generations.

Tips for Interviewing Prospective Employees

Prepare Structured Questions: This helps you manage consistency in evaluating all candidates.

Ask Behavioral Questions: This may reveal how a candidate handled past situations. In hiring, past performance is a likely indicator of future performance.

Assess Cultural Fit: The vital importance of your company's culture is discussed in Chapter 2 and must be taken into account in evaluating how a prospective employee might fit in that regard.

Devise Skill Tests: Certain positions within a company lend themselves to short tests. You or your office manager can devise a few questions to assess a candidate's qualifications for a certain clerical position. Your bookkeeper or CPA may do the same regarding bookkeeping positions.

100 | *12 Hiring the Right People*

Conduct Panel Interviews: After a positive private interview with a prospective employee for a position, invite others to sit in for an additional interview. This provides multiple perspectives on the candidate's suitability.

Evaluate Communication Skills: All of your employees should have the skills to clearly express ideas and collaborate effectively.

Ask About Problem-solving: Ask the candidate to give an example that demonstrates his effectiveness in managing through an unusually challenging work situation. Ask him to describe a case in which he tried and failed. His analysis could be a clue about how he accepts responsibility.

Provide Examples of the Job Responsibilities: The better the candidate understands the job before accepting an offer, the more likely he will succeed.

Check References: This is a tricky task because many employers do not provide more than confirmation of and dates of employment, if even that. But you may get enough feedback to verify the accuracy of some of the information on the candidate's application. When you do get the chance to speak with a *cooperative* former employer, be prepared to ask specific questions and follow-ups to assess the former employer's honesty. Former employers are often reluctant to speak freely because of friendship with their former colleague or for fear of personal or legal repercussions.

The final test remaining is your own judgment. You cannot go through the above screening without forming an independent opinion. Even in light of the input you get from others in your company, it is you who pulls the trigger. Especially when hiring a key person, say your accountant, whose competency, reliability, and honesty are crucial, you'd better feel good about it. No contractor has never hired the wrong person for the job at hand, but with due diligence and sound judgment you will win much more often than you make the wrong decision.

Job Benefits

Here are a few benefits that job candidates are known to value highly, and to look elsewhere for employment if they are not offered. Those you make available to your employees will be driven by your company's unique circumstances.

Remote Work for Office Workers: Options for remote work or hybrid work arrangements.

Recognition: A sense of being needed and appreciated.

Flexible Hours: Ability to have flexible working hours to maintain work–life balance. Responsible employees should be expected to consider their impact on the company when exercising this flexibility. It may not be practical for employees whose presence is critical to the company, but compensating alternatives can be explored.

Modern Tools: Use of up-to-date technology and tools for efficient and effective work.

Digital Communication: Platforms for instant messaging, video conferencing, and collaborative work.

Safe and Respectful Environment: Policies and practices that ensure a safe and respectful workplace for all employees, both office and jobsite.

Training and Development: Access to continuous learning opportunities and professional development.

Clear Career Path: Defined pathways for career advancement and growth: Clear understanding of what is required for career advancement.

Mentorship Programs: Opportunities for mentorship and guidance from experienced professionals.

Meaningful Work: Engaging in work that has a positive impact and aligns with personal values. The chance to make a difference. Finding purpose in their work.

Corporate Social Responsibility: Working for companies that are committed to social and environmental responsibility.

Open Communication: A culture of open and transparent communication from management.

Feedback Culture: Regular and constructive feedback to help employees grow and learn.

Fair Pay: Competitive salaries that reflect the skills and contributions of employees coupled with timely compensation reviews. Fair pay is critical to attract and retain the employees you need for sustained growth and profitability.

Bonuses and Incentives: Performance-based bonuses and incentives.

Health and Wellness: Health insurance, wellness programs, and mental health support.

Retirement Plans: Access to retirement savings plans such as 401(k) and financial planning resources.

Paid Time Off: Reasonable paid time off policies, including vacation, sick leave, and parental leave.

Work–Life Integration: Policies that support work–life integration, such as family leave and childcare support.

Trust and Responsibility: Trust in employees' abilities to manage their tasks and responsibilities independently.

Empowerment: Empowering employees to take initiative and make decisions.

Innovative Culture: A culture that encourages innovation, creativity, out-of-the-box thinking, intellectual stimulation, and fulfillment.

Integrity: Leadership that demonstrates integrity and ethical behavior.

Company Character: The company's makeup, character, reputation, and standing in its industry.

Transparency: Transparency in decision-making and business practices.

Environmental Responsibility: Commitment to sustainability and environmental responsibility.

Community Engagement: Active engagement and contribution to the community.

It is your choice in most cases whether or not to provide these benefits and conditions to your employees but most will be demanded by the more recent generations of job seekers in the workplace.

Jobsite Facilities and Conveniences

Your jobsite superintendents' needs for outdoor construction projects are of course different than office workers', and it is important to provide facilities, conveniences, and safety features that enable them to perform their duties effectively and safely. Here are some key elements to consider, which will vary widely depending on project size, complexity, and duration.

Office Trailer: One or more mobile office trailers equipped with drawing racks and desktops, chairs, filing cabinets, and necessary office supplies. This serves as a base for the superintendent to conduct meetings, manage paperwork, and coordinate site activities.

Internet and Communication: Reliable internet access, phones, fax machines, and radios to ensure seamless communication with the project team, owner representatives, and subcontractors.

Jobsite Facilities and Conveniences | **103**

Computer and Printer: A computer with project management software, critical path software, and a printer/scanner for handling documentation, blueprints, and reports.

Restroom Facilities: Accessible restroom facilities, either in the form of portable toilets or permanent structures, depending on the duration of the project.

Break Area: A small space maintained for amenities like a refrigerator, microwave, and coffee maker.

Storage Space: Secure storage for personal belongings, tools, equipment, and important documents. This may require an additional trailer or other space.

Personal Protective Equipment (PPE): Ensure that the superintendent has access to necessary PPE, such as hard hats, safety glasses, gloves, high-visibility vests, and steel-toed boots.

First Aid Supplies: A well-stocked first aid kit for treating minor injuries on site.

Emergency Plan and Equipment: Clear emergency plans and procedures, along with equipment like fire extinguishers, emergency eyewash stations, and defibrillators.

Site Security: Measures to secure the construction site, including fencing, security cameras, lighting, and potentially security personnel to prevent unauthorized access and protect against theft and vandalism.

Safety Signage: Appropriate signage to indicate hazards, safety instructions, emergency exits, and first aid locations.

Training and Meetings: Regular safety training sessions and toolbox talks to keep the superintendent and all site personnel informed about safety protocols and updates.

Proper Lighting: Adequate lighting for both day and night operations to ensure visibility and reduce the risk of accidents.

Sanitation and Cleanliness: Maintaining a clean and organized site to minimize hazards and create a safe working environment.

Health and Wellness Resources: Access to health and wellness resources, including water stations and rest areas.

If you are a small, startup contractor, your jobs may not need some of these provisions.

12 Hiring the Right People

Onboarding New Employees

This section will have more or less importance for you depending on the size and scope of your business.

According to the Society for Human Resource Management (SHRM), replacing an employee costs a company an average of six to nine months' worth of the employee's salary. For an employee earning $100,000 annually, this translates to recruiting and training expenses ranging from $50,000 to $75,000. Those numbers may not hold in your small business but in scale the dollar cost, time consumption, and disruption can be significant.

Thus, minimizing employee turnover is good for your bottom line. Start with an effective onboarding process for new hires. There are several things you can do during onboarding to lay the foundation for your new hires' success.

Load them up with information about their new company. Get all of the new employee forms and procedures out of the way. Introduce them around the office and to the field employees they'll be working with.

These procedures are more important if you are hiring from the workforce whose methods of learning and other factors limit the skills they need for productivity, working with others, addressing bosses and customers, and greeting and interacting with vendors and visitors.

Social skills and business protocol they will need when talking with customers and vendors in person or on the phone may need to be explained, along with appropriate use of email and text messages.

There have been reports in the business media that strategies are sometimes necessary to instill a work ethic in some younger workers, including being productive and taking initiative. An employee's strength in sensing need, taking timely action, and following through to resolution in a given situation can be as impactful as, say, winning or losing a bid. Or foreseeing shortage of plywood you need for a job you're in the middle of and scouting out alternative sources before the shortage hits. Or in preventing an accident. A critical employee skill that I call *see need and take action* has long been a factor in my hiring, training, or appraising the value of an employee to my company.

Talent First, a workforce development outfit based in Grand Rapids, Michigan, advises businesses to hire based on a willingness to learn rather than on skills. This would not work when hiring for positions involving science, technology, engineering, and mathematics, i.e. STEM positions, but it makes sense for a construction firm that hires other workers entering the workforce.

Employee Meetings

In many workplaces, performance reviews have evolved from the traditional annual assessments to practically continuous feedback sessions. Many employees want regular check-ins to receive timely feedback on their performance that helps them meet their goals and be engaged with their work. This approach can boost performance and promote a culture of ongoing development and adaptability.

But let's say you employ a crusty job superintendent who began his construction career a few decades ago. He wants to know that you are pleased with his work, but he gathers that information directly or indirectly every time you or his project manager visits his jobsite. A good word or a pat on the back now and then. He wants to feel satisfied that he is fairly compensated, respected, and appreciated. He looks for strength and confidence in you. Many of the men who worked for me – my project managers and field superintendents – would work day and night to get a job done when the proverbial *ox was in the ditch*, and there were many such occasions over the years. Such men are of grit, of loyalty, of love of their work, and of pride. One of the many ways you maintain this loyalty is by never letting an employee, subcontractor, or vendor come directly to you in circumvention of the employee whose direction they are expected to follow unless that employee accompanies them.

Employee Handbook

The very exercise of creating an employee handbook forces you to think through issues that will come up and to establish policies and procedures to accommodate them. Do this before your first hire or you will inevitably make commitments contrary to the policies you establish later, and you will have to live with them. You may use generic handbook templates as guidelines that you can conform to policies and procedures that apply to your unique circumstances and preferences. You can always expand and improve them in employee favor, but it can be difficult or practically impossible to change or take back something valued by your employees.

Please see an employee handbook example in Appendix F.

Noncompete Agreement

A noncompete agreement is a legal contract between an employer and an employee or independent contractor that restricts the latter from engaging in certain competitive activities for a specified period after the termination of their employment or contract.

Proper drafting of a noncompete agreement requires a lawyer who is current on US as well as state laws regulating employment practices. Following are practical and general aspects of a noncompete agreement you as a business owner should know, but they should not be used except under the advice of your attorney.

Noncompete agreements are designed to protect the employer's legitimate business interests, such as trade secrets, client relationships, and proprietary information. But to be enforceable, noncompete agreements must be carefully drafted to meet specific legal criteria, as courts often scrutinize them to ensure they do not unfairly restrict an individual's ability to earn a living. A generic noncompete agreement typically includes several essential elements.

A noncompete agreement must clearly define the scope of restricted activities, specifying the types of businesses, services, or geographic areas that the employee is prohibited from engaging in. For instance, a noncompete might intend to prevent a former employee from working for a direct competitor within a 50-mile radius for two years after leaving the company.

The agreement should specify the duration of the noncompete restriction. This period must be reasonable since overly long restrictions are often deemed unenforceable by courts. Periods ranging from six months to two years are considered reasonable duration in some cases, but this varies based on the industry, jurisdiction, and other case-specific factors.

The geographic boundaries of the noncompete must be defined. It should only restrict the employee from competing in areas where the employer has a legitimate business presence or interests. A nationwide or global restriction might be considered excessive unless the employer operates on such a scale.

The agreement must include something of value provided to the employee in exchange for agreeing to the noncompete. This could be in the form of continued employment, a promotion, or a financial benefit.

In an attempt to ensure that a noncompete agreement will withstand potential legal challenges, your lawyer will likely take several factors into consideration.

Tailoring the noncompete clauses to the specific employee's role, responsibilities, and access to sensitive information makes the agreement more likely to be seen as reasonable and necessary by a court.

Courts are more likely to enforce a noncompete if the duration and geographic scope are reasonable and directly related to protecting the employer's business interests. Overly broad or lengthy restrictions can render the agreement unenforceable.

A severability clause allows the court to remove or modify any unenforceable parts of the agreement while keeping the rest intact. This increases the chances that the noncompete will be partially enforceable even if some provisions are deemed excessive.

Noncompete laws vary significantly by state and the agreement must comply with the specific laws and public policies of the jurisdiction where it will be enforced.

Adequate consideration or value to the employee, fairness, and reasonableness will increase the likelihood of enforceability. If an employee is asked to sign a noncompete after starting employment, additional consideration may be necessary.

Please see Appendix B for an example of noncompete agreement.

13

Banking

A bank is a place that will lend you money if you can prove that you don't need it.

Will Rogers, American
humorist, actor, and social
commentator

The sudden failure of the Silicon Valley Bank in Santa Clara, California, in 2023 triggered depositors' concern for small banks. Deposits up to $250,000 are currently guaranteed by the US Government Federal Deposit Insurance Corporation and, respecting that limit, your small business will likely fare better by choosing a small or regional bank rather than banking giants such as Bank of America or Wells Fargo. If your deposits exceed the guaranteed limit, you can continue the FDIC protection by following exacting rules that must be complied with to ensure your deposits are guaranteed. Contact the FDIC directly for any clarification of the complex rules.

The giant banks are less interested in relatively smaller loans and deposits that are more often associated with small businesses, and their service and attention to your needs may receive a lesser priority than you would like. They also move their employees around to different positions and locations more than smaller banks, making it less likely that you can establish and grow a beneficial ongoing relationship with your contacts at the bank.

By contrast, local or regional banks are highly dependent on small business deposits and loans and will genuinely value your business. You want a banking relationship in which your bank officer becomes an expert about you, your company, and your goals. Smaller banks give you a chance to know and form

Mastering the Construction Startup: A Business Infrastructure Guide, First Edition. Nick B. Ganaway.
© 2025 John Wiley & Sons, Inc. Published 2025 by John Wiley & Sons, Inc.

110 | *13 Banking*

ongoing relationships with higher-level employees who have more interest and authority in fulfilling your financial needs.

If you need to take out a loan or if you have a temporary cash flow crunch, the regional or local bank manager or loan officer who knows you and your company will not need to start over and relearn your history every time you meet with him or her. And you are more likely to get to know the people at the top in a small bank, meet for lunch occasionally or on the golf course – things that oil the gears of business. Such relationships are practically guaranteed to make your banking more personally and professionally rewarding.

It's important to establish relationships with more than one bank. Having loans and deposits in the same bank gives the bank too much control over you. For example, if you have a delay making loan payments, the bank could freeze your operating accounts, limiting your ability to work through a temporary cash-flow problem.

Another reason to do business with more than a single bank is that if one of your banks goes out of business or becomes part of a giant bank through merger or acquisition, you won't be left without an established bank relationship.

Loan Documentation

If you apply for a loan, you will be required to provide your bank or other lender with several forms of documentation. A professional, organized, and well-thought-out presentation is certain to improve your chances for loan approval. Numbers are by no means the only factor that your banker will take into account when considering your request.

Write your loan request in simple, clear language describing the amount you are requesting and what the loan will be used for. Include information about your education, experience, specialized knowledge, and other characteristics important to starting and operating your business.

If you are already in business, the bank will require your business and personal financial statements for the past three years. Financial statements include balance sheet, income statement, and cash flow projections. If you are not familiar with the information comprising each of the components of your financial statements, you will do well to learn. Refer to the brief descriptions of financial statements in Chapter 11 of this book.

Identify the collateral you offer and its dollar value. Be aware that if the assets representing the collateral decline in value during the term of the loan, you will be required to replace that value.

Personal Guarantee

Your lender is likely to ask you to personally guarantee any loan. Do not do this without considering the potential consequences. Personally guaranteeing any loan would put you at risk of losing all of your business and personal assets, including equipment, bank accounts, real estate, stocks and bonds, automobiles, and so on if you are in default. The lender may be willing to limit your guarantee to a specified dollar value or to certain assets that you own. If this is not possible, you should carefully weigh what is at stake before agreeing to this requirement.

Variable Rate Loans

Lenders often offer the option for a variable rate loan that carries an enticing lower initial interest rate than a fixed rate loan. As many developers and home owners have sadly learned, a floating-rate loan puts you subject to an unpredictable economy and the mercy of lenders. Consider the risk before opting for a variable rate loan, especially a long-term loan.

For some time preceding the COVID-19 pandemic that began in early 2020, interest rates in the United States were stable and many borrowers took out loans bearing floating interest rates. With floating rate loans, the starting rate is firm for a set period of time, say five years, and then adjusts automatically at every fifth-year anniversary for the life of the loan. Economic inflation ensued in the aftermath of the pandemic and increased interest rates by several percentage points, increasing the loan payment accordingly. Many borrowers with these often-huge loans were not able to make the higher payments and still maintain positive cash flow, and many forfeited the mortgaged property, sold it at a loss, or went bankrupt.

Credit Unions

You may find that using a local credit union meets your particular needs better than a bank. Credit unions typically offer lower interest rates and fees, provide more personalized service, and focus on building long-term relationships. Your choice between banks and credit unions depends on your particular circumstances.

112 | 13 Banking

Private Lenders

Private lenders can be an option for borrowers who do not qualify for traditional bank or credit union loans, but it's important to take into account the potential disadvantages. Higher costs, less regulation, and greater risks associated with private lending can outweigh the benefits, making it crucial for borrowers to thoroughly research and compare all available options before making a decision.

Private loans usually have higher interest rates, origination fees, processing fees, and miscellaneous charges that may greatly increase your borrowing cost. Less regulation means possibly less protection for borrowers.

Private lenders typically offer shorter loan terms, leading to higher monthly payments, and require substantial collateral to secure the loan. They often don't have the same level of reputation or trust as established banks and credit unions and probably are less transparent about the terms and conditions of their loans.

These lenders typically focus solely on lending, meaning borrowers may miss out on additional support and services, such as bank accounts. More importantly, they may be less willing to work with borrowers who face difficulties in repayment.

Private lenders might include family members or friends, but this should be considered only after all other options are ruled out. The risk of damaging an important relationship is high and possibly permanent.

Please see Appendix C for a sample business loan proposal.

14

Construction Insurance

Insurance is the DNA of the economy. It provides the necessary security for businesses to take risks and innovate, ensuring stability and peace of mind in society.

> Warren Buffett, American business magnate and one of the world's most successful and influential investors

Commercial insurance companies offer coverages far beyond the scope of this book to serve the needs of all kinds of business operations, but here are a few basics. To be sure you understand and acquire the coverages appropriate for your company, contact a reputable commercial insurance agent who will assess your insurance needs based on your particular circumstances. Construction-related insurance, of which there are many forms, has its own provisions and requirements.

Upon establishment of your insurance coverage, your agency will issue you a certificate of insurance (COI), stating coverage dates, the coverages you bought, the insurance carrier(s), the amount of coverage, and other important aspects of your policy.

Although the COI shows the details of your coverage and is very likely reliable, you can rely only on the terms stated in your policy or policies. Upon your receipt of the COI, verify that all information on the form is as you expected and complete. In the unlikely event it is not, notify the agent immediately, preferably before starting the work covered by the policy.

Set up a fail-safe system that will alert you to approaching critical policy dates to ensure against overlooking them.

Mastering the Construction Startup: A Business Infrastructure Guide, First Edition. Nick B. Ganaway.
© 2025 John Wiley & Sons, Inc. Published 2025 by John Wiley & Sons, Inc.

14 Construction Insurance

To be sure you are holding a current and valid COI for a business you are doing business with, accept it only from the business's agency, not from the business itself. This will avoid your being blindsided by lapsed or canceled policies. Having provided you the COI, the agency should notify you of policy changes, but that is not certain. You can contact the carrier directly to verify coverage.

This is a lesson I learned only too late, at considerable expense to my company. Years ago, an employee of one of my subcontractors was seriously injured on our jobsite. When my office notified the sub's insurance agency, we were told that the policy had been canceled. This meant that all medical and other costs related to this injury would hit my company's WC policy, which affected my WC premium for several years going forward.

In this case, we had carelessly accepted the sub's insurance certificate directly from the subcontractor's office rather than the sub's agency. By following our own rules, we likely would have been notified of the cancellation in time to deal with it before damage was done.

If you were to need to add another party as additionally insured under your policy, e.g. your banker, you will need to have the carrier issue an endorsement to the original policy, naming the banker as an additional insured party. Merely requesting the agent to issue a COI naming the banker as an additional insured is not sufficient.

I tend to be over-insured for the peace of mind it provides.

It can pay you generously to shop around with other reputable agencies before your insurance policy renewals to ensure against your agency or carrier becoming less competitive.

Following are brief general descriptions of several of the more common types of insurance many businesses need. Each type can be customized to your requirements. These descriptions lack critical detail but may give you information that will help you discuss your insurance needs with your agent.

General/Commercial Liability Insurance

General Liability insurance (CGL) coverage protects your business against financial loss that results from, among other things, bodily injury, property damage, medical expenses, and judgments, and provides for the defense of claims and lawsuits against your company by others. If someone is injured on your jobsite and files a claim or lawsuit against your company, your CGL coverage will take over, investigate the claim, and handle it to its conclusion.

Workers' Compensation Insurance

If you are a business owner with employees other than yourself, you will fortunately or unfortunately become very familiar with workers' compensation (WC) insurance.

So, what is WC insurance?

> "Workers' compensation is a form of insurance that protects a business owner from claims by employees who experience a work-related injury or illness – either sustained on business premises or due to business operations. Typically, workers' compensation covers the employee's medical expenses, rehabilitation costs and at least some portion of their lost wages. If an employee is killed on the job, it also pays a funeral benefit."
>
> *(National Association of Insurance Commissioners)*

WC insurance is a critical component of a comprehensive business insurance plan, particularly in industries like construction where workplace injuries are more common than in many kinds of business.

WC is generally a no-fault system, meaning that employees can receive benefits without having to prove that the employer was at fault for their injury. The primary aim is to ensure that injured workers receive medical care and compensation for lost wages without undue delay and without the need for legal intervention.

Here are key features of Workers' Compensation Insurance

Cost of Medical Treatment: WC covers the cost of medical treatment necessary to diagnose and treat the injury or illness. This can include emergency room visits, surgeries, medications, rehabilitation services, and any other medical expenses related to the injury.

Income Replacement: If an employee is unable to work due to a job-related injury or illness, WC provides partial wage replacement. The amount and duration of these benefits can vary, but they generally cover a portion of the employee's average weekly wage.

Disability Benefits: These benefits compensate for lost wages if the employee is temporarily or permanently unable to return to work.

Rehabilitation Benefits: WC may also cover the cost of vocational rehabilitation services to help the injured employee return to work. This can include job training, career counseling, and assistance with job placement.

Death Benefits: In the unfortunate event that a work-related injury or illness results in an employee's death, WC provides death benefits to the employee's dependents. These benefits typically cover funeral expenses and provide financial support to the deceased employee's family.

Employer Responsibilities Under Workers Comp Insurance

Employers are required to carry WC in most states. Failure to do so can result in severe penalties, including fines and potential legal action. Employers must also report work-related injuries and illnesses promptly to their insurance carrier and comply with any state-specific WC regulations.

Workers' Comp Insurance Benefits to Employers

While WC insurance primarily benefits employees, it also offers significant advantages to employers. It helps mitigate the financial impact of workplace injuries by covering medical and legal expenses, thereby protecting the business from potential lawsuits.

Cost of Workers' Comp Insurance

Your cost for WC depends on the kind of business involved. As you would expect, the rates for construction workers are higher than rates for office workers.

Your cost is adjusted up or down based on the dollar cost of the claims your company incurs in the previous period or periods. This system is designed to encourage employers to establish and enforce rules that promote safe workplaces. It is definitely in your financial interest to do so.

You can dispute an employee's WC claim if you believe their injury or illness was not work related. In some cases, you may deem it wise to hire an attorney who specializes in WC claims, as WC is a very specialized field involving the employer, the injured employee, and the company's insurance carrier.

In summary, workers' compensation insurance is a vital safeguard for both employees and employers. It ensures that injured workers receive necessary medical care and financial support while protecting businesses from the financial and legal repercussions of workplace injuries.

Builders' Risk Insurance

The owner of a construction project requires the contractor to obtain builders' risk coverage to protect against loss or damage to the property, such as by fire or weather.

While the construction project is underway, be aware of any potential increase in the value of the project. If events, even the passage of time, cause the value of the insured property to increase you must get the policy carrier's written endorsement of the change in order to maintain protection under the policy. Failure to do so may result in you becoming coinsurer.

This is because if your *policy amount* becomes less than 80% of the *insured property value*, you as the insured become a coinsurer with the carrier and responsible for a portion of the loss *even if the actual total loss is less than the policy amount*. Co-insurance is more thoroughly explained in the following text.

Business Owner's Policy

Your agent may determine that a Business Owner's Policy (BOP) is well suited for your small business.

A BOP is a comprehensive insurance package that combines property and liability coverage into a single policy. This type of insurance is particularly popular among small- and medium-sized businesses. By bundling property and liability insurance, BOP insurance provides coverage for claims related to bodily injury, property damage, and other liabilities that businesses may face. It offers a convenient and cost-effective solution for protecting the assets and interests of various smaller businesses.

Product Liability Insurance

This coverage can protect your company against financial loss resulting from a defective product your company manufactures, sells, or distributes that causes injury or bodily harm.

Professional Liability Insurance

For businesses or individuals that provide professional services to customers, this coverage can protect against financial loss resulting from errors, omissions, negligence, or malpractice.

Commercial Property Insurance

This coverage protects your business against loss and damage of company property resulting from fire, smoke, civil disobedience, wind, hail, and vandalism.

Fidelity Insurance

You can buy a fidelity bond or insurance that offers you protection against losses caused by employees' fraudulent or dishonest actions such as embezzlement or establishment of phantom vendors. A phantom vendor situation is where a company employee establishes a fictitious vendor in the company. Then the fraudulent employee submits false invoices from his fake business to your company. The invoice is routinely routed through the company's payables process, approved by the fraudulent employee, and paid by your company to the fictitious vendor owned by the employee. This is not an uncommon occurrence in businesses.

In a phantom vendor case I am familiar with, the contractor discovered such a situation without letting the dishonest employee know, quietly arranged for law enforcement to come to his crowded work area during office hours, place him in handcuffs, and march him out as his fellow employees watched in shock and disbelief. (If the contractor was trying to set an example for other employees who might ever be so inclined, he probably accomplished that.)

If you plan to obtain fidelity insurance for an employee(s) or for employees in a certain job category, it is best to do so upon hiring the employee as a routine matter. You can imagine that unexpectedly doing so later may signal to the employee that he is not trusted whether that is the case or not.

Home-based Business Insurance

This coverage is very limited, but if you run your business out of your personal home, you may be able to add it to your homeowner's insurance policy as a rider.

Coinsurance

Property insurance policies require the policyholder to insure a certain percentage of the property's value. If the insured value is less than the required percentage, the policyholder becomes a coinsurer along with the carrier and must bear a portion of a loss. This potentially devastating situation occurs if the value of the insured property increases due to, for example, inflationary pressure, improvements made to the property, new real estate development in the area, or other reasons.

Here is an example. An insured property was originally valued at $5 million when the insurance policy was taken out. The policy included an 80% coinsurance clause, meaning the policy amount was required to be at least 80% of the property's value. Several years later, a fire caused $1 million in damage. However, due to rising real estate values, the insured property's value had increased to $8 million at the time of the loss, but the owner had failed to adjust the policy amount to reflect the increased value.

These are the calculations used in determining the insurance award to the policyholder.

- Amount of insurance actually carried: $5,000,000
- Required insurance coverage based on new value: $8,000,000 × 80% = $6,400,000
- Amount of insurance carried/required insurance coverage = $5,000,000/$6,400,000 ≈ 0.78125%
- Loss due to fire: $1,000,000
- Insurance award: $1,000,000 × 0.78125 = $781,250 (minus any deductible)

The policyholder would only receive $781,250 for the $1,000,000 loss due to the coinsurance penalty.

In this scenario, the property value increased significantly after the initial policy date, but the owner did not adjust the insurance coverage accordingly. As a result, the coinsurance clause penalized the policyholder, reducing the insurance payout for the loss. It underscores the importance of regularly reviewing and adjusting insurance coverage to reflect current property values to avoid such penalties.

To avoid becoming subject to a coinsurer with your carrier, you can check into the availability of a policy "rider" that should increase the coverage amount automatically.

Making Claims

Your insurance carrier is an indispensable business resource to your company but don't make the mistake of thinking the friendly agent you deal with, or the carrier, will generously help you beyond a strict interpretation of your coverage and its analysis of the circumstances underlying your claim, which are often not a black and white issue. In cases of coverage uncertainty, expect the carrier to fight for the advantage. To be sure that your claim is settled fairly and properly, consider getting your attorney involved in initially reporting the claim to avoid inadvertently compromising your position. It is all too easy to make a seemingly harmless, likely recorded comment to the insurer when reporting a claim that may later be used against you.

Performance and Payment Bonds

In the construction industry, performance and payment bonds are crucial for ensuring that projects are completed as per agreement without related undue financial distress. By providing financial protection and security to project owners, contractors, subcontractors, and suppliers these bonds help maintain the integrity and viability of construction projects. They are the carriers' guarantee to the insured that the contractor will perform his contractual duties (performance bonds) and pay all his cost incurred in doing so (payment bonds), thus playing a vital role in the management and successful execution of construction contracts.

Performance and payment bonds are issued by insurance companies to the contractor who has a certain level of net worth and personally guarantees any loss covered by the bonds. I have warned against personal guarantees elsewhere in this book but when it comes to performance and payment bonds, you may not have a choice. You may be able to limit your exposure to only certain parts of your personal net worth.

The better news is that some owners will waive the requirement for performance and payment bonds for contractors they use repetitively, after gaining a comfort level with them.

Performance Bonds

Performance bonds and payment bonds often go hand in hand. A performance bond is issued to ensure that the contractor completes the project according to the specifications laid out in the contract. If the contractor fails to complete the project satisfactorily, the bond will cover the costs to complete the project or repair deficiencies. The aim is to financially protect the project owner against losses that would occur if the contractor were to default.

The process typically involves three parties: the principal (contractor), the obligee (project owner), and the surety (the company who issues the bond guaranteeing the contractor's performance). If the contractor defaults, the surety has the option either to finance the contractor to completion, bring in a new contractor to complete the project, or pay the project owner the bond's value.

Payment Bonds

A payment bond is a type of surety bond posted by a contractor to guarantee that the subcontractors and suppliers are paid for the work and materials provided. This bond is crucial in protecting the supply chain in construction projects, ensuring that those who have provided labor and materials to the contractor will receive compensation even if the contractor goes bankrupt or fails to pay.

Similar to performance bonds, payment bonds involve the contractor as the principal, the project owner as the obligee, and the surety. Payment bonds are particularly important because they prevent project delays due to disputes between subcontractors and suppliers who might otherwise place a mechanic's lien on the property if they are not paid.

Summary

Proper insurance coverage protects both your personal and business assets against unexpected events. To ensure that you have the right coverage, deal with a reputable, qualified business insurance agent or agency who knows construction insurance and its unique requirements. Be aware that you are likely at a disadvantage when communicating regarding claims with an insurance professional whose job is to minimize its exposure in any claim. Establish procedures to ensure that you do not become coinsurer. Examine initial and renewal certificates of insurance to be certain that they are consistent with the contractual requirements.

The ACORD certificate of insurance form is used to summarize essential information about an insurance policy. While there are ACORD forms for different types of insurance, one commonly used in the construction industry is the Certificate of Liability Insurance.

Please see Appendix I for an ACORD certificate of insurance example.

15

Business Plan

If you fail to plan, you are planning to fail.
Benjamin Franklin, Founding Father and Entrepreneur

Purpose and Evaluation

The value of and need to objectively research and evaluate the conditions relative to the services you plan to offer cannot be overstated. These evaluations prepare you to logically and clearly lay out your qualifications and those of your key employees, your access to adequate capital, and your commitment to proper business management. This is the basis of your business plan.

In developing your business plan, think of it as a go/no-go fact-finding project essential to your success. *You must be willing to modify or even shelve the plan if your research doesn't support it.* That's a very hard decision, but changing course now to seek an *A*-level opportunity is exponentially better than taking an unwarranted risk with your original plan that is sub A-level at best. Don't put the cart before the horse by implementing your preliminary ideas before following the above rules.

Presentation for the User

What follows your above research is to organize your business plan with the perspective of the lender or others who will need it in mind.

Mastering the Construction Startup: A Business Infrastructure Guide, First Edition. Nick B. Ganaway.
© 2025 John Wiley & Sons, Inc. Published 2025 by John Wiley & Sons, Inc.

15 Business Plan

There is no universal format for writing a business plan, but here are some important elements.

- Estimate your cost and time to get to the point of opening the doors to your new company and after that your expected operating expenses until the company can pay them.
- Show how you determined the best marketplace for your product or service.
- Describe your access to capital.
- Show how your company will make money.
- Describe your competition and how your company will compete.
- Explain how your company will gain a competitive advantage over other contractors.
- List your business connections and strategic relationships.
- Show how your unique services will benefit your customer, such as your personal or professional expertise, any special training, company management depth and experience, your capitalization, and any intellectual property.
- Describe your company's operation – how it will interact with your customers.
- Explain your banking relationships.
- Be concise but thorough.
- Include additional information that is unique to your proposed business.

Proposal Guidelines

Your business plan is filled with all of the necessary information, but the way it is presented is what powers it. Business protocol is that it should be typed on 8.5" × 11.5" white or off-white stationery using black traditional font. Spelling and grammar should be correct, and the documents should be logically organized in a simple binder and, if helpful, indexed.

The arrangement of your business plan should be flexible. If you are seeking a bank loan or line of credit, for example, strategically position and emphasize the information that you believe will be of greatest interest to that bank. Clearly state the amount you are seeking early in the letter.

Your audience, whether one person or a roomful, will form its opinion of you and your company from the whole of your presentation. This includes your confidence in yourself and your ideas; any casual or pertinent remarks you make; the organization and appropriateness of the written materials or other documentation you offer; your personal affect including the way you are dressed, and your demeanor, attitude, and more.

The mention of religion and politics should always be avoided. Assume that everyone in the room has different opinions than yours about any subject, so yours is best left unsaid. Never use foul language orally or in written materials. If your host starts the meeting with casual conversation, perhaps to ease any tension of the moment, you should not consider it a good time to tell your favorite joke. Let your host set the tone and take the lead until he or she turns it over to you to make your presentation.

Speak directly to the person you're presenting your proposal to in steady voice while making comfortable eye contact. If you are seated at a table or desk, sit with your back straight or lean in to explain your materials, signifying engagement with them and the person you're dealing with. Avoid becoming defensive if your host probes into some element of your materials. Just matter-of-factly explain.

When the meeting is over express appreciation for the opportunity you were given. If given the chance spend a moment with each person in the meeting. Follow up the next day with a short note of thanks to the host.

You and I might agree that some of these characteristics are unrelated to the business at hand and should play no role in any business decision, but the decision makers you will be dealing with are human beings, not algorithms. It is in your interest to take that into account.

If you are a businessman or woman who is experienced in such circumstances, these suggestions may seem over the top. But the reader who may be unaccustomed to protocol in such a setting is sure to find it useful.

Please see the business plan examples in the appendices.

Part 5

Ideas

16

Niche Contracting

Success in business is all about focus and doing what you do best. Niche contractors exemplify this by mastering their specific domains.

Richard Branson, British entrepreneur and
founder of the Virgin Group

In 1974, McDonald's had 2,272 outlets in the United States and about 3,000 system-wide. By March 2024, there were 13,529 US locations and about 41,800 worldwide. Such growth, which seems to have no end, is a gift to general contractors who want to specialize. Overall, around 80% of my company's revenue came from fast-food construction and continues today beyond my ownership.

Advantages of Niche Contracting

Opportunities for niche contractors are abundant and visible. Beyond just new establishments, a steady stream of replacement and renovation projects in areas of growth become available to the general contractor. Every few years, retail businesses upgrade their facilities to stay competitive, presenting continual demand for construction services. Chain store operators who regularly build, rebuild, or upgrade their facilities often stick with one or very few general contractors who specialize in fast-food projects and have gained favor by way of competitive pricing, above-standard quality and service, effective working relationships between the parties, and turning over projects on the promised schedule.

My construction niche being the fast-food restaurant industry, I will explain niche contracting using that industry as an example. However, many regional

Mastering the Construction Startup: A Business Infrastructure Guide, First Edition. Nick B. Ganaway.
© 2025 John Wiley & Sons, Inc. Published 2025 by John Wiley & Sons, Inc.

and national companies in other industries operate on chain store or franchise basis and fit the same mold as fast-food restaurants.

The potential for specialization is demonstrated by data from the National Restaurant Association: The restaurant and foodservice industry added 300,000 jobs in 2023 for a total of 12.37 million workers in about 750,000 restaurants.

Convenience stores, often paired with gas stations, are prime candidates for chain store contractors. Typically backed by financially sound major oil companies, these projects also follow residential and commercial growth patterns.

The renovation or rebuild market offers just as much potential for contractors as the new-store segment. Chain store operators maintain modern, attractive appearances to *keep up with the Joneses* and to address the wear and tear on their facilities. Remodel jobs are typically more profitable per dollar for contractors than new construction.

In addition to periodically scheduled upgrades to their stores, restaurants frequently roll out new products that require new behind-the-counter food-preparation equipment and electrical and mechanical modifications, all of which may require extensive work in small spaces after closing hours, running up the cost. A general contractor who is awarded a contract to make such revisions consecutively to multiple existing stores benefits from the learning curve. As an example, in the early days of restaurant drive-through windows, restaurant chains not only began including them in their building designs but went on a binge adding them to their existing restaurants. Efficiencies resulting from the learning curve effect can result in a better bottom line for contractors who are awarded multiple consecutive projects. The contractor's cost for the second project in a series of like-kind jobs is less than the first one, and so on, ultimately leveling out at a more profitable bottom line.

Restaurant construction is complicated by the large amount of mechanical and electrical work required in relatively tight spaces, stringent finish standards, compressed construction schedules, and the various trade workers sharing crowded workspaces. Not all contractors and subcontractors are well-suited to such fast-paced, densely packed projects. I have seen large, major-project general contractors try fast-food construction, thinking it's good fill-in work during down economic periods, but then move on after learning that their setup and their field personnel are not geared for small jobs.

The outlook is promising for general contractors specializing in chain store construction. While specific requirements and nuances vary, the fundamental characteristics are largely consistent and compare favorably to the wider commercial construction market. I always believed we had less competition than nonspecialized generals in the broader market. Fast-food companies locate construction managers in cities strategic to their operations, who put contracts out for bids and coordinate with the contractors. They usually find it to their advantage to use familiar contractors as described above. They require less input from the owner representatives and can bid lean because of the commonality among fast-food projects.

Construction contracts within the chain store market span a wide range, from several hundred thousand dollars to multimillion-dollar projects. Within this niche there is flexibility to operate as a small-scale general contractor or to expand into somewhat larger operations.

Additional Niche Advantages

There is a certain amount of commonality among fast-food chains' building designs. They are usually built on one-half to three-acre sites. Electrical and mechanical systems are roughly the same. Construction methods are more or less uniform and interior finish materials are typical. Site work on such small square footage, often one-half to two acres, is easy to price and accomplish.

Subcontractors sometimes specialize as well, and those who focus on fast foods or similar projects give generals a leg-up in finding qualified subcontractors. This saves time in bid preparation and yields more uniform subprices.

I have never built for a fast-food chain or franchisee who wasn't in a mad rush to get their business into operation. This is understandable because their cash flow is outward instead of inward until they open for business. Interest is accruing on their bank loans. They may have key staff on the payroll.

Beware the ambitious owner's construction manager who doesn't understand or respect the contracting business but wants to make a name for themselves. We had built for the Wendy's chain for several years when they brought in a new representative to hire contractors and put jobs out for bid. He announced to the Wendy's bidders including my company that instead of comparing the bottom line of all bids to select the apparent bid winner, he would compare the lowest price *on each line-item* of each contractor's bid sheet and add up those select prices to create his own fictitious total. He would then hold up that price to the initial low bidder for him to accept or reject the job on that price basis. This doesn't work.

16 Niche Contracting

Many line items in a contractor's bid proposal encompass more than a single category of cost and it's unlikely that any two bidders will allocate those charges the same way. Take heating, ventilation and air conditioning (HVAC) for example. Bidder A may include the HVAC electrical work in the HVAC line item, while Bidder B includes it in Electrical. The bids from both A and B have the work covered, although in different line items, and so on throughout the line items.

You can see how cherry-picking line items results in nonrepresentative numbers. We no longer submitted bids on Wendy's projects until its bid analysis process returned to normal.

By specializing as a general contractor focused on one of the subsegments within chain store construction, say, fast-food restaurants or automotive parts and repair stores, you unlock the potential for a consistent stream of profitable projects with relatively lower associated risk.

Throughout my years building fast-food restaurants, the competitive landscape did not change much. We often won negotiated contracts with repeat customers, but even in more-common competitive bidding two to four bidders was the norm, and they were usually contractors we had competed with.

Avoid competing in lowest-bid projects when possible and strive toward developing relationships that lead to negotiated work. In private and public sectors alike, lowest-bidder jobs can breed adversarial relationships and result in low-profit margins for contractors and a never-ending chase for work. It is a tough business for contractors who get no negotiated work.

Some chain store operators use in-house architectural and engineering departments for designing their own buildings but use outside architects and engineers to adapt the company's stock designs and specifications to meet local codes and site conditions.

The specializing general contractor has the opportunity to establish a cost database specific to a given chain of stores, refining it with every subsequent job. This database is invaluable when bidding on future projects, enabling bid tweaking to the bare minimum when deemed necessary.

The construction industry is here to stay. The rise of technology and even artificial intelligence are little threat to construction industry jobs. The building and maintenance of structures will always require on-site work by plumbers, electricians, HVAC companies, grading contractors, carpenters, roofers, and various other building trade professionals. Construction firm owners, managers, superintendents, and other key personnel will continue to play vital roles.

Industry experts express concerns about the insufficient number of qualified candidates entering construction to meet the demand for managerial and leadership positions. The industry is ripe for men and women looking for lucrative and rewarding careers.

Chain Store Owners and Franchisors

A vast number of chain stores operate in the United States and typically use free-standing buildings. They are categorized as Quick Serve Restaurants (QSRs), Casual Dining Restaurants, Automotive Stores and Other Chain Stores. These companies most often use free-standing buildings.

Quick-Serve Restaurants

- McDonald's
- Burger King
- Wendy's
- Taco Bell
- Chick-fil-A
- KFC
- Subway
- Arby's
- Sonic Drive-In
- Popeyes
- Jack in the Box
- Dairy Queen
- Hardee's
- Carl's Jr.
- Little Caesars
- Domino's Pizza
- Pizza Hut
- Papa John's
- Raising Cane's
- Culver's

16 Niche Contracting

Casual Dining Restaurants

- Applebee's
- Chili's
- Olive Garden
- Red Lobster
- Outback Steakhouse
- LongHorn Steakhouse
- Texas Roadhouse
- TGI Fridays
- Buffalo Wild Wings
- Red Robin
- Cracker Barrel
- IHOP
- Denny's
- Ruby Tuesday
- Carrabba's Italian Grill
- P.F. Chang's
- The Cheesecake Factory
- Bob Evans
- Golden Corral
- Black Bear Diner

Automotive Stores

- AutoZone
- Advance Auto Parts
- O'Reilly Auto Parts
- Pep Boys
- NAPA Auto Parts
- Carquest
- Firestone Complete Auto Care
- Goodyear Auto Service
- Jiffy Lube
- Midas
- Valvoline Instant Oil Change
- Meineke Car Care Center
- Discount Tire

Chain Store Owners and Franchisors | **135**

- Tires Plus
- Big O Tires
- NTB – National Tire and Battery
- Les Schwab Tire Centers
- AAA Car Care Plus
- Tire Kingdom
- Monro Auto Service and Tire Centers

Other Chain Stores

- Walmart
- Target
- Lowe's
- The Home Depot
- Costco Wholesale
- Sam's Club
- Best Buy
- Walgreens
- CVS Pharmacy
- Rite Aid
- Tractor Supply Co.
- Dollar General
- Family Dollar
- Dollar Tree
- Big Lots
- PetSmart
- Petco
- Hobby Lobby
- Michaels
- Office Depot
- OfficeMax
- Staples
- Ace Hardware
- True Value
- Sherwin-Williams
- Menards
- Harbor Freight Tools

16 Niche Contracting

- The UPS Store
- FedEx Office
- AT&T Store
- Verizon Wireless
- T-Mobile Store
- Sprint Store
- Aaron's
- Rent-A-Center
- GameStop
- Mattress Firm
- Rooms To Go
- Ashley HomeStore
- Bed Bath & Beyond
- 7-Eleven
- Circle K
- Wawa
- Sheetz
- QuikTrip
- Speedway
- Casey's General Store
- Cumberland Farms
- Maverik
- Love's Travel Stops & Country Stores
- RaceTrac
- Pilot Flying J
- Kum & Go
- Kwik Trip
- QuickChek
- GetGo
- Stewart's Shops
- Rutter's
- Royal Farms
- Stripes Convenience Stores

17

Outside Board of Advisors

We all need people who will give us feedback. That's how we improve.
                        Bill Gates, co-founder of Microsoft

Limited liability companies, sole proprietorships, and partnerships are not required to have a formal board of directors, but as the owner of your small business, you may decide to look into establishing one. An advisory board is typically made up of outside experts who lend their knowledge and guidance to an organization to help it grow, avoid pitfalls, and achieve its objectives.

Unlike a formal corporate board of directors, members of a board of advisors have no fiduciary responsibility and are not held liable for the company's actions. They have no authority within the company and, for example, cannot make hiring and firing decisions.

There are several reasons you might consider having a board of advisors. The cliché, *It's lonely at the top*, soon proves itself to be true. Other than your spouse or other person close to you, whom would you bring into your inner sanctum? A golfing buddy? Someone you used to work with? The dentist who lives next door?

Not likely. That may be because you think no one else can possibly understand what you're going through, you are not comfortable sharing your financial matters with others, you think if you open up to someone they may take it as an invitation to give you unsolicited – and unhelpful – advice, or you may be protective of certain business secrets. Maybe you're a private person by nature. Every owner has his or her own unique reasons for keeping things close to the vest.

Carefully selected by you, a board of advisors offers a way around that. The members you choose are likely to be businesspeople you think will be strong

Mastering the Construction Startup: A Business Infrastructure Guide, First Edition. Nick B. Ganaway.
© 2025 John Wiley & Sons, Inc. Published 2025 by John Wiley & Sons, Inc.

17 Outside Board of Advisors

in areas you are not proficient in. They may have a managerial background or sales experience, technical know-how, or accounting, or law – or any fields that are not your strong suit. Seldom does a new business owner have deep experience in every aspect of running a business. As hard as it may be to admit, you may benefit from a little outside help.

There are many accounts in biographies and various business publications about idea-men and -women who brought their startup baby to life but knew or found out the hard way that they lack the particular skills, perhaps, or the experience to guide it to ultimate maturity. They either hire those who do meet these requirements or sell the business to another outfit.

You would seek out potential board members who meet certain needs. To ensure that you get fresh and objective advice from members, don't include your lawyer or accountant or other people in your business circle. Your advisory board members need to see things from a clean perspective and be able to offer unfettered ideas.

It's generally not a good idea to bring family members or friends onto your advisory board. Sometimes, it is necessary to remove or replace a board member and if he is your brother-in-law it may become an emotional event or worse.

A contractual relationship may be established with board members. You can set term limits that allow you to rotate members whose skills you may no longer need and bring in others as your business grows or changes course.

Startups may offer a small equity in the company or pay in cash based on time involved and the value a member brings to the table. Contract templates are available online.

Is an Advisory Board Necessary?

Whether to establish a board of advisors is a personal choice and it doesn't have to be decided initially.

Some businesspeople believe that an entrepreneurial startup needs an advisory board, whose members individually and as a group will help you cross some of the startup barriers, give you guidance, and point out areas of risk as you grow to the next level. A member with the right background may be able to prevent problems or take advantage of opportunities that you might miss.

One or more members of your board may prove to be a valuable contact and resource long after their board association has ended. And, as said, having a board can be partial relief to owner isolation.

Many business owners go it alone. I was happily one of those but in retrospect an advisory board might have raised questions that would have guided me around costly management mistakes I made over the years.

If you want to avoid the time and expense of recruiting and maintaining a board of advisors, you may know and trust a few people you can call up or meet for lunch and talk about things weighing on your mind. All the better if they have business experience. People you can talk to as friends, not as consultants. That is worth considering.

Please see Appendix E for a board of advisors agreement example.

18

Case Study of an Existential Business Crisis – A Personal Account

Every adversity, every failure, every heartache carries with it the seed of an equal or greater benefit.
> Napoleon Hill, American self-help author best known for his book *Think and Grow Rich*, and his focus on personal success and the principles of achievement

The following is an analysis of the decisions and conditions that led to a cluster of problems occurring practically in sync that presented an existential threat to the author's company.

Many years ago, my 12-year-old general contracting firm was coming off a banner year, well established with a number of national customers, substantial cash reserves, enthusiasm among the managers and staff, and a track record of successful years since its beginning.

This positive set of circumstances gave me confidence to move forward with the expansion I'd been thinking about for some time and to implement it in the current year.

The expansion involved opening a satellite office in Orlando, Florida, more than 400 miles from my Atlanta headquarters, which required renovation, new furnishings, and additional computer systems. Adding the necessary managerial, administrative, and accounting staff would bring the total company payroll to about 40 employees. I met with my key managers and laid out my idea and reasoning. We kicked it around over the next weeks and none of them voiced any concerns (more about that later.) I greenlighted the project.

The project required substantial capital and increased overhead expenses. It also focused my and my employees' attention on this project rather than on

---

Mastering the Construction Startup: A Business Infrastructure Guide, First Edition. Nick B. Ganaway.
© 2025 John Wiley & Sons, Inc. Published 2025 by John Wiley & Sons, Inc.

18 Case Study of an Existential Business Crisis – A Personal Account

our customers, and it resulted in a layer of management between me and my front-line employees.

While the dust was still settling on these challenges, three of my national customers either filed for bankruptcy or discontinued operations in states we did business in – all within a few months of each other. These three customers had accounted for about a quarter of my company's profit in the prior year.

Another Shoe Drops

As my company was reacting to this problem, it came time for our annual audit by our outside CPA firm.

The auditors uncovered a large financial loss for the year, which had not shown up in the internal reports I was seeing each month. Of course, if they had been accurate and timely, I would not have implemented the expansion project, and all the expense and disruption that accompanied it would have been postponed or eliminated.

I stressed in earlier chapters the importance of hiring excellent employees and this episode emphasizes the point. The company's bank accounts had not been balanced across several months and my in-house accountant accepted responsibility for this egregious omission. Mine was the biggest accounting department she'd managed in terms of complexity, revenue, and the number of people she managed. She had become overwhelmed but kept it to herself, thinking she could work her way through it.

But the responsibility was mine. I should have recognized that the company had outgrown her and at some point asked more questions, being more tuned in to the stress she was under. To avoid being caught with my pants down going forward, I determined a few telltale numbers that I wanted on my desk every Friday. I called this my *Vital Signs Report*. Although limited in what such a cursory report can convey, it gave me enough information to ask revealing timely questions.

A Perfect Storm

It wasn't over. In what became a further setback, my bank upon receiving the new audited financials promptly canceled the line of credit we used intermittently and killed the chance to obtain the loans I would need to manage through this situation, a scenario that has been the death knell for many a small business. The unfortunate irony here is that a year earlier I was lured away from the bank I'd used for many years. That happened when one of my primary contacts at my original bank took a job at a competitive bank and offered me generous concessions to change. My former bank, where I had forged relationships at all levels over a number of years, would likely have hung in with me and provided the loans that would have made our climb easier out of this deep hole. As it was, our ability to continue business as usual was under pressure.

The loss of revenue from my three customers was a problem within itself, but clearly the impact of the expansion decisions I made early in the year made the situation much worse.

Based primarily on the good results of my company's prior year, I had bet on the unreliable expectation that the next year would be as good as the last and absorb the expansion's cost.

Larger, well-established corporations can take a substantial hit in stride. They have deep pockets to absorb a substantial loss, and they are not likely to have a quarter of their revenue disappear over a few months.

Looking Back

Most of my managers were graduates from Georgia Tech University, and young, smart, and ambitious. I probably had never expressed that I expected them to voice their opinions if different from mine when important decisions were under discussion, and in this one, deciding whether to open a satellite office along with all its ramifications, they did not. If I had given them freedom of speech, so to speak, there is a pretty good chance that important questions would have been floated regarding the risk factors involved – including cost, timing, even the need for a second office at all.

Participation with the boss in making bigger decisions is the opportunity for young managers to learn how decisions are arrived at, how to be part of them, and to make a difference in outcomes. My failure to have given my guys the permission, or rather the *responsibility*, to jump in when they had an idea or concern deprived me of their potentially pivotal input and denied them important growth experience.

A Tough Decision

With the loss of the company's reserves, I was faced with lending the company a substantial chunk of my personal assets outside of the company in order to continue operating. Even a cash infusion was no promise of a successful outcome. Failure then would be a blow to the wherewithal I would need to start over.

On the positive side of things, I had the goodwill and dedication of my employees and a long record of good business practices with our customers and vendors. I drew heavily on these precious intangible assets to string out payables for 60–90 days.

To keep things going, my managers and new in-house accountant met weekly to decide how to divide the scarce cash on hand among the most urgent payables. Fortunately, the company had no debt (other than to me, for the cash I put in to keep us going.) Some of my key employees offered to take temporary cuts in pay and I completely suspended mine.

The Personal Effect

What few others knew during this period was the private hell I went through for more than two years that felt like 10. I went to sleep and woke up every morning knowing that a single additional burp in my company or any one of its underpinnings could bring it down. And since I had risked much of my personal savings as the only hope to save the company, it would be much harder to start over if it failed.

While my employees knew we were in crisis mode, only I had the big picture, the overall responsibility, and the painful consciousness of what was at stake. Sincere friends and professionals will give you moral support, speak encouraging platitudes, pat you on the shoulder, but it is not possible for anyone, even your wife or husband or your most-valued employee to put themselves into your state of mind in such a situation and grasp bone deep the anguish, the risk at hand, the loneliness, your concern for your employees, your worry for your family – the *personal* crisis you're going through. This is the essence of the cliché, *It's lonely at the top*.

This is what every business owner who has risked everything to make a go of it faces when things go south. And many have faced such situations. All businesses start small and if such a crisis occurs before they have built up cash reserves, solid banking relationships, loyal customers, or an experienced management team, it is then that they are most vulnerable. But, if you live in fear of failure, you're not likely to reach your potential.

I had started running 10 years earlier for its health benefits. Now it became as essential as the food I ate – both for physical maintenance and the brief mental relief it provided from relentless pressure. One of the rewards of running is the practically worry-proof cocoon it wraps you in, as all runners know, for about an hour after the run, which I would spend on my patio swing. A lifesaving break. I must have been miserable for my wife and kids to live with, but they provided respite when I was at home in the evenings. They had the advantage of not knowing the dire potential of the situation.

Success, at Last

It took two grueling years that tested my and my employees' nerves, commitment, and perseverance to stop the bleeding and once more return to profitability – a good month followed by a bad one and so on, rarely a clear windshield. Through it all, my company never missed making the payroll even in those toughest days, no vendor went unpaid, and no employees left despite the extra burden imposed on them.

The loyalty and grit demonstrated by my employees is something that cannot be trained or bought, and I have unending gratitude for them and for their rare kind. I no longer own the company but I am fortunate long since those days to be able to enjoy regular contact with many of them who live nearby and to call them my friends.

My wish for you, fellow entrepreneur, is that you will never find yourself in the circumstances I've described, but if you do, I hope your recollection of this story will contribute to your confidence and perseverance.

When you're in the middle of this kind of situation, it's impossible to think, *Wow, this is really good for me!* But succeed or fail, there is inevitable positive fallout. This *resilience* – i.e. our ability to face the most extreme of hard times in our lives while at the same time coping with our education, our jobs, difficult bosses, impossible red tape, our marriages, raising our children, making ends meet, and on and on – this *resilience*, this durableness we've earned that allows us to face it all *cannot and does not come from experiencing only smooth sailing throughout our lives. It can only come from struggle, hard work, times of*

18 Case Study of an Existential Business Crisis – A Personal Account

uncertainty, relentless pressure, and the worry that comes with such risk. This is how character is built.

I might claim that there were some circumstances beyond my control, such as the sudden loss of key customers, but part of my job in running the business was to be knowledgeable and wide-eyed enough to anticipate pitfalls and to have things in place to deal with them. It's not realistically possible to plan for an asteroid striking Earth, or even a once-in-a-century pandemic, but loss of customers and vicious economic cycles are nothing new and must be included in prudent business planning.

I've had to think for a minute what I'd have done if still another bomb had dropped in the midst of our scramble to recover. I cannot imagine that I would have said, *Well, that's it* and shut down. There is almost always another door, another avenue. Many biographies and autobiographies of hugely successful people would never have been written had it not been for their preceding failure that taught them valuable lessons. Think Henry Ford, Steve Jobs, Walt Disney, to name a few.

Never give up when there's a lot at stake. If you succeed, you will be a better person for it and even if you fail you will be a better person for it in many ways, and well ahead of the game the next time around. The three men named above would tell you so.

An undeniable factor in any enduring success is that of luck. Anyone who has succeeded in a long-term venture in which a level of risk and uncertainty is involved and takes all the credit for it without acknowledging that luck played a role is either fooling himself or trying to fool you.

Success has many parents. Effective relationships, enthusiasm, curiosity, knowledge, determination, commitment, motivation, desire, raw intelligence, time, place, circumstance, and luck work together in one beautiful combination.

I recount this personal history in hopes it is a relatable example of America's capitalist system in which each of us is allowed to succeed and allowed to fail. This freedom to fail is one place where the government is least intrusive.

It is also a story of employee hardship, grit, loyalty, determined team effort, and ultimate return to profitability. A reminder that bad decisions can be made in good times, and often are.

19

Investing in Real Estate as a Parallel Business

Ninety percent of all millionaires become so through owning real estate.
Andrew Carnegie, Scottish-American industrialist and
philanthropist who led the expansion of the American steel
industry in the late nineteenth century and became
one of the richest individuals in history

As your construction business grows and becomes profitable, you're likely to find yourself with disposable income to invest.

Your goal for these invested funds is not to generate current income but to ultimately create wealth without taking your time and focus away from your contracting business.

With that in mind, I would be shortchanging you if I didn't tell you about an investment vehicle I have used since my early entrepreneurial years called *single-tenant net-lease* (STNL) real estate ownership. STNL properties are often purchased and owned by professionals like doctors and lawyers and successful entrepreneurs such as general contractors or subcontractors running their businesses.

These are people who often have funds they would like to invest but not the time, nor often the knowledge, to seek out suitable investment opportunities. Of course, there is always the stock market, but I will explain later the comparative advantages of net-lease investments.

Mastering the Construction Startup: A Business Infrastructure Guide, First Edition. Nick B. Ganaway.
© 2025 John Wiley & Sons, Inc. Published 2025 by John Wiley & Sons, Inc.

Owning STNL Properties – a Successful Proven Strategy

Any CVS, Walgreen's, Dollar General, AutoZone, Starbucks, Chase Bank, Home Depot or practically any major brand you visit or pass on the street every day may well be STNL. Most often, a man or woman no different from you or me owns the property and receives the rent check every month.

In simplest terms, a typical net-lease deal is one in which, say, the operator/owner of a Taco Bell restaurant (the company) builds its building and sells the building and ground to an investor like you or me. The investor then leases it back to the company in a *sale-leaseback* (SLB) transaction. Thus, the company becomes the investor's tenant and pays the investor rent for a period of 10, 15, 20, 50, or more years, as agreed to by the company and investor and specified in the lease.

The investor is not involved in the business operation.

What defines this arrangement as a single-tenant net-lease deal is that the company pays the three basic costs of real estate ownership: property taxes, maintenance, and insurance. Therefore, the rent received by the investor is *net*, or free, of those expenses associated with the property.

Investment in STNL properties is at the center of a subindustry within the larger real estate industry.

General Characteristics of STNL Properties

For this discussion, the term STNL incorporates triple-net lease properties (NNN) and sale-leaseback properties.

The typical company in an STNL arrangement is the franchisor of a restaurant, car repair business, convenience store, or any of the chains listed in Chapter 16. To facilitate an SLB transaction, the company pays for all financing, testing, research, permitting, purchase of the building site, and for creating the plans and specifications that will be used for the project and, finally, paying a building contractor to build the building and all improvements.

Once the facility is ready to be put into service, the company puts the property on the market. This is where the investor comes in. Once the company accepts an investor's offer, the investor makes his own arrangements for the purchase. Thorough investigation of the property (known as "due diligence") by the investor is essential, which may be largely facilitated by his attorney.

As you can see as the potential investor, the only significant time this acquisition takes you from your primary business occurs at the front end, and less so at the back end. You must establish an LLC in which to hold the property as described elsewhere in this book. Only you, of course, can make the critical decisions about the viability of the company, the property location, the lease details, and the financing prior to committing to the deal, but your attorney and real estate broker will do much of the legwork and lay out those factors for you to consider. Note that the company's real estate broker does not represent your interest in the process but can provide many of the due diligence materials that you and your attorney request.

As a general contractor for numerous restaurant chains, I had contracts to build the facilities for the owner and upon completion of a given project I, personally as the investor, often had opportunities to purchase the property from the company and simultaneously lease it back to them – the sale-leaseback deal. This is a hand-in-glove fit. More often than not, the contractor who built the premises does not become the investor/landlord to the company/tenant.

The STNL lease document legally lays out the terms agreed to by the company and the investor. This might be a 20 to 30-page document detailing the rights, privileges, duties, and responsibilities of each party, the start and end date of the lease, the rent amount, and all other pertinent details. The lease controls the property.

If you are the investor, your attorney will hold your hand through the transaction and ensure your legal protection by way of the lease. When the deal closes, your work is essentially done for the duration of the lease, although for your protection you must verify every year (or policy period) that the tenant has paid the property taxes and provides you with proof of insurance coverage. And that you have received the monthly rent payment. Once the lease is signed by both the company and the investor it cannot be changed for the duration of the lease except by agreement between the parties. As the investor in the property, you will want to visit it periodically to see that the tenant/company is maintaining the property as required in the lease. The reasonable expense incurred in visiting your property for inspection is tax deductible. If it is located in a desirable place to visit, so much the better.

Many companies use this business model in order to recover most of their development costs through the SLB method and use the recovered capital to develop more of their operating locations. The company pays the rent to the investor out of its operating income. A well-researched and implemented deal is a win–win situation for both company and investor.

150 | *19 Investing in Real Estate as a Parallel Business*

A less common variation of NNN is an NN lease, in which the investor is responsible for some part of the costs of ownership. The discussion in this chapter refers to the NNN lease.

The STNL Marketplace

Many well-respected national and regional real estate brokerage firms have a division that specializes in for-sale STNLs and publishes lists along with their descriptions. You can sign up for the free emailed listings. They are useful to the investor, who uses them to compare offerings, to watch for trends in cap rates, and to make an offer on an STNL property of interest. STNL property listings are easily found online and often advertised as "triple-net" or "NNN" properties.

A property listing usually provides the three primary factors involved in determining the value of an STNL property: the net operating income (the rent paid to the investor), the purchase price, and the capitalization rate (cap rate). The cap rate is the percentage return the investor receives on his investment in the property. If any two of the three primary factors are known, the third can be determined with simple algebra. This tool can be used to weigh alternatives to the numbers presented.

Cap rate = net operating income / purchase price

Purchase price = net operating income / cap rate

Net operating income = purchase price × cap rate

In most offerings, the net operating income is a fixed amount. By using the formulas, the prospective investor can play around with the cap rate or the purchase price in Excel to arrive at a dollar amount he or she is willing to offer. If price negotiations follow, the formulas become even more useful for analysis.

On the buyer side of the STNL coin are investors who specialize in STNLs. In a broad sense, the brokerages and investors know how to find each other and they do, creating a thriving, popular marketplace for this niche investment vehicle.

Primary Factors in Pricing an STNL Property

If you ask a real estate professional the three most important factors in determining the value of a piece of real estate, the answer you will get is *location location location*. No doubt you've heard this.

Location of course is not the only important criterion, but it is almost always the first filter in your search for a suitable property. If someone offers you a building or tract of land in an area you know to be depressed, declining, crime-ridden, etc., you probably won't consider it further.

But among the offers you will see, one good location may be better than another good location. Location then becomes only one of the factors you will consider in evaluating a property, any one of which may not meet your criteria.

For example, dollar store real estate such as Dollar General has been a pretty hot net-lease investment for several years. The company/tenant is financially very strong, making it attractive to the investor. But dollar stores are often located on the fringe of small, rural communities with marginal prospects for future growth. If you own such a property not located near or among other businesses that create traffic and shopping activities and the company leaves at the end of a 15-year lease, your chances for re-leasing or selling your vacant building profitably are not ideal unless the surrounding area has developed during the lease term. I have taken a pass on dollar store investment for these reasons.

An ideal prospective STNL property must meet several criteria. It is located in a prominent location, in an economically healthy market area with positive demographics for the future, and in close proximity to other major brands. It will be located near stable residential areas and have good visibility, easy ingress and egress, and on a heavily travelled two-way street.

A property that meets all those criteria is rare so the variables must be weighed. If you are considering a property with a thriving ongoing business with a good track record to lease back to a company, that takes a lot of the guesswork out of the equation.

Google Earth street view offers a convenient way to look at a property you are interested in by clicking along the street to view the nearby businesses. You can use the aerial view to see commercial centers, where the business centers start, where they end. If you are scouting out a property whose listing caught

19 Investing in Real Estate as a Parallel Business

your eye, a short investigation on Google Earth might give you enough information to decide against it, saving you a trip to look at it, or it might gin up your interest.

While Google Earth is preliminarily very helpful, I have never bought a property without going to see it. I want to watch the traffic flow and check how well-kept the property and nearby properties are, both inside and out. I want to get a sense of the backup neighborhood, go inside the property, sit down, and order something to eat if it is a restaurant. I want to compare the property location with other business areas nearby.

Is there a newer town center to which customers are likely to migrate? Do most of the properties on the street seem to be well maintained? Are all of the nearby retail places occupied and operating? Are For Sale signs popping up everywhere? Do I get a sense that the property location will still be thriving when the lease ends in 15 years? Is the city vibrant? Do published demographics point to continuing growth?

As you have observed, consumer stores more or less flock together. You seldom see a drug store, for example, standing alone far from other retail businesses because shoppers and buyers tend to go to denser marketplaces for one-stop shopping. Thus, the prevalence of strip shopping centers and malls.

A top consideration in choosing a net-lease property is the financial muscle backing up the lease. This is based on a number of factors including the net worth of the company you lease to, its brand value and recognition, the number of units it owns, and, in some cases, environmental factors. You and your attorney or real estate agent can usually get a pretty good handle on this question. The general expectation is that the parent company is more financially substantial than one of its franchisees, but there are many very strong franchisees that are equally secure to invest in and can be even safer.

Many financially strong multi-location companies don't franchise, meaning they are operated by the company. These include CVS, Home Depot, AutoZone, Walgreen's, and Dollar General, to name a few. But that doesn't mean they don't do SLBs. Their financial strength is reflected in the relatively lower percentage return an investor is willing to take in a net-lease deal with such super-strong companies. Lower risk means lower return on investment.

Subject to the criteria described in this chapter, STNL deals are considered a conservative investment, and many buyers pay cash or trade into them through a tax-free exchange ("1031") described below.

If you require a bank loan for all or part of the property's purchase price, interest rates and types of loans come into play. Ideally, the rent income you receive from the company each month will cover the loan payment.

But that may not always be possible, and you will have to supplement the rent income with money from your reserves or from your primary business operation. Remember that this is an investment, just as if you were buying companies listed on stock exchanges, not a business expense you can write off on your taxes. Your bank loan should have a *fixed* interest rate. A low-interest floating rate loan can be tempting but when the rate changes, so do loan payments, thus changing your cash flow situation. This scenario has led to the downfall of investors all too often.

If the rent income for a net-lease deal you're considering will not cover the loan payment in the beginning, take into consideration your main operating company's prospects for future growth and profitability. In addition, most net-leases include automatic annual or periodic rent increases to help offset inflation. Given that it is an investment in your future, you may decide that the supplemental payment is justifiable. Few entrepreneurs have succeeded without making sacrifices and taking risks.

Your reward for any burden your net-lease investment puts on you is that the bank loan, if there is one, will one day be paid off – maybe 10, 15, or 20 years – leaving you with the entire rent income at your disposal. This will go on for the duration of the initial lease and possibly beyond. It can be your retirement income, achieved with relatively little worry, work, and expense, and with a prudent level of risk. However, like any investment nothing is guaranteed.

Replacing the Property Early

Within about a year before the lease is set to expire, the investor should revisit the lease to prepare next steps. STNL leases usually give the company the option, but not the obligation, to extend the lease for a designated number of years by notifying the investor of its intention a specified number of months in advance. If the company extends, there is usually nothing for the investor to do at this point.

However, if the investor receives notice from the company that it will not exercise its option to extend the lease, the investor must begin preparations to find a new tenant or to sell the property. If the property and demographics have

19 Investing in Real Estate as a Parallel Business

remained favorable, the investor will probably be happy if the company does not or cannot extend the lease, giving the investor the opportunity to increase the rent for a new tenant. A prominent STNL property surrounded by other thriving businesses is a gift that keeps on giving to the investor.

Every company uses its own criteria to determine whether to extend the lease, but if the demographics and other conditions affecting the value of your property have remained constant or better since the beginning, there is a good chance the company will take the option to continue the lease.

If for some reason the investor needs or prefers to sell his property mid-term of the lease, there is a thriving market for STNLs having 10 remaining lease years or even fewer, although their value may gradually drop over time when the remaining term of the lease falls below 10 years.

STNL Properties Versus the Stock Market

Busy professionals and other businesspeople who have a flow of income above their routine expenditures often turn to a stockbroker or asset manager who puts their money in stocks and bonds. However, brokerage, management, and other fees and expenses will reduce that return by as much as 2 or 3 percentage points, or more.

An example of how volatile, worrisome, and devastating the stock market can be is the US "Great Recession" of 2007–2009. The unemployment rate rose from 5 to 10% and the Dow-Jones industrial average fell 54% in just 17 months. The average investor who stayed in the market to the bottom lost half of his or her investment on paper in a year and a half. Those who bailed out of the market somewhere on its way down missed its eventual recovery. Those who braved the storm to the bottom of the market and held on saw their stocks recover and far more over the following years. But it was a stressful and uncertain time for everyone in the market.

Banks and businesses failed at a high rate. Many small businessmen and women lost their businesses and part or all of their life savings and yet remained responsible for an outstanding bank loan balance. Half-built residential and commercial developments stopped dead in their tracks. Home values fell by double digits for the first time since the Great Depression of 80 years earlier. Homebuilders cratered. There was no way anyone could be certain the economy and the stock market would recover, and most individual citizens were in a state of panic.

While I was among those who were worried sick about the stock market, my rent income from my STNL properties continued unabated. This was also the case during the 2020–2023 Pandemic.

The Case for Investing in Real Estate

For many investors, including myself, it is much more rewarding and less stressful to watch STNL rent payments regularly hit our bank accounts on the first of every month than to watch stock market values gyrate from one day to the next and remain only as a paper asset until ultimately sold and generates a tax liability.

The rent payments deposited into your bank accounts can be used to pay the mortgage on the property, if there is one, or be immediately reinvested for additional returns. If your property is located in another city or state, you can travel there for periodic "inspection" of your property and write off your travel expenses against its income.

If you own real estate you can stand on it, you can see it and you can feel it and touch it. You can sit and look at it and take pictures of it. You can go inside and walk around. That may not sound like much, but it is, as you will see if you buy and own an STNL. Chosen by informed process, it will serve you well.

I do not mean to pour cold water on the stock market. On the contrary owning both stocks and real estate gives you needed diversification and helps balance your investments. But I find owning STNL properties much more exciting.

Dealing with Capital Gains Tax

Selling your property at any time subjects you to capital gains tax if you sell for an amount above your cost in the property.

However, if you take the defensive position of selling your STNL property mid-term of the lease, or at any time for that matter, you may decide to buy a replacement property and start the cycle over. If so, you can take advantage of an Internal Revenue Service (IRS) provision by which any capital gains tax resulting from the sale can be delayed and in certain cases, avoided. This is routinely done by using the proceeds from the sale of the original STNL to buy a replacement or other real estate by way of the IRS tax provision well known to commercial real estate investors as a 1031 tax-deferred exchange or simply a 1031.

19 Investing in Real Estate as a Parallel Business

This valuable tool, which can produce wealth for you and your heirs, is the icing on the real estate cake.

The 1031 Tax-deferred Exchange

The 1031 is applicable to many kinds of assets but this chapter describes it only as applicable to STNL properties.

Simply put, this IRS provision allows an investor to sell a net-lease property, immediately use the proceeds to buy another property "of like kind" and defer any taxes generated by the sale. Both capital gains and recaptured depreciation taxes are deferred. The tax burden shifts to the "replacement" property and becomes due only when the replacement property is sold. This can be done serially, kicking the tax bill further down the road each time.

Upon your death, by law, the replacement property will get a step-up in basis to its present value for tax purposes. This means that if your heirs sell the inherited property, they will pay capital gains taxes on only the gain since your death, not the gain since the original purchase. This is a very valuable tax advantage.

If you buy your first net-lease property at, say, age 30, it is possible to accrue wealth over your lifetime by buying and holding or exchanging STNL properties essentially tax-free and passing it on to your heirs. It is passive investment except during the process of acquiring or exchanging your property, allowing you to focus your time and energy on your construction business.

STNL real estate is an enjoyable wealth-building strategy. You have no employees to manage and almost no overhead costs. You don't have to find a new tenant every year or two. You have no clogged toilets at 3 a.m. or roof leaks to deal with and there are tax advantages.

Strict rules must be followed when initiating a 1031 exchange and some qualified professionals specialize in them to make the transaction relatively easy for the investor.

In Conclusion

While passive investments are not strictly infrastructure for your main operating businesses, they are very much infrastructure for your successful entrepreneurial *career*. From personal experience, I recommend STNL investments as an exciting opportunity that is both manageable for the busy entrepreneur and also builds wealth for you to draw on in later years or pass on to your heirs.

Here is a caveat worth noting: If you require long-term financing to buy the property under consideration, you may want to jump in only if you can expect to pay the mortgage off by the time you might sell your company or retire, and if you also can manage any monthly mortgage payment amount that exceeds the rent income. You of course will be building equity in the property but will not have positive cash flow until the loan is paid off, which may make the deal less appealing. The older you will be when the loan is paid off, the shorter time the deal will put cash into your pocket.

20

Summary Checklist for Startup Businesses

What to Do First

- Establish initial meetings with your CPA and business lawyer to decide the corporate legal entity best suited to your circumstances, i.e. LLC or other, among other preliminaries.
- Meet with an insurance agent to discuss and apply for your insurance needs.
- Register your corporation or LLC with the secretary of the state where your headquarters is located, and with each other state in which you will do business.
- Open your business bank account(s).
- Each LLC or corporation should have its individual bank account.
- Obtain any applicable professional licenses.
- Obtain a business license.
- Obtain an Employer Identification Number for your corporation or LLC at IRS.gov. If you plan to operate as a sole proprietorship, your Social Security Number is used for this purpose.
- Buy and implement financial and accounting software systems.
- Determine policies and procedures for accounts receivable, payables, tracking expenses, and cash outlays, and designate a certain employee responsible for this financial function, if not yourself. Make this effective from Day 1.
- If you have one or more employees, establish a system to track and pay payroll taxes when due. This is not a flexible deadline.
- Establish vendor accounts.
- If you have multiple businesses, establish separate LLCs or other corporate entities for each one so that claims or suits against one will not put the assets of the others at risk.

Mastering the Construction Startup: A Business Infrastructure Guide, First Edition. Nick B. Ganaway.
© 2025 John Wiley & Sons, Inc. Published 2025 by John Wiley & Sons, Inc.

160 | *20 Summary Checklist for Startup Businesses*

- Maintain strict separation of your personal and business finances. Do not commingle funds. Sign all business forms and documents in the business name along with the signee's title (owner, president, etc.)
- Pay bills when due to establish credit and reputation.
- Create written and signed agreements for your business transactions, which at a minimum identify the parties to the agreement and date; specify the product or service to be provided; establish performance dates; state amounts payable or receivable including due dates; and incorporate the legal requirements for contracts.
- Avoid signing unlimited personal guarantees.

21

Useful Reading

Here are a few construction industry and general business periodical publications.

Construction Dive

Provides in-depth journalism and insight into the most impactful news and trends shaping the construction and building industry. The daily email newsletter and website cover topics such as commercial building, residential building, green building, design, deals, regulations, and more: www.constructiondive.com.

Construction Business Owner

A magazine focusing on the business side of construction, including management, finance, and operations: www.constructionbusinessowner.com.

Journal of Light Construction

A practical publication for residential and light commercial contractors, covering techniques, tools, and business advice: www.jlconline.com.

Qualified Remodeler

A magazine dedicated to remodeling professionals, featuring market trends, business strategies, and project case studies: www.qualifiedremodeler.com.

Mastering the Construction Startup: A Business Infrastructure Guide, First Edition. Nick B. Ganaway.
© 2025 John Wiley & Sons, Inc. Published 2025 by John Wiley & Sons, Inc.

Construction Executive

A publication offering business news, strategies, and insights for senior-level construction executives: www.constructionexec.com.

Engineering News-Record

A leading publication providing news, analysis, and data for the construction industry worldwide: www.enr.com.

Entrepreneur

A magazine that provides insights, advice, and news on small business management, startups, and entrepreneurial endeavors: www.entrepreneur.com.

Inc.

A publication focusing on growing companies, offering advice, tools, and services to help business owners and entrepreneurs: succeed: www.inc.com.

Money

A personal finance magazine that offers advice on saving, investing, and managing money, helping readers make smart financial decisions: www.money.com.

Forbes

A leading business magazine known for its lists and rankings, covering finance, industry, investing, and marketing topics: www.forbes.com.

Wired

A magazine that covers technology, science, culture, and business, providing in-depth analysis and insight into the future of these fields: www.wired.com.

The New York Times

A major American newspaper that offers comprehensive news coverage, opinion pieces, and features on a wide range of topics, including politics, business, technology, and culture: www.nytimes.com.

The Wall Street Journal

A major American newspaper that covers national and international news with a focus on business: www.wsj.com.

In addition, a few timeless books I've valued throughout my business career and personal life are as follows:

How to Win Friends & Influence People by Dale Carnegie
Think and Grow Rich by Napoleon Hill
Emotional Intelligence by Daniel Goleman, PhD
Make Your Bed by Admiral William H. McRaven
See You at the Top by Zig Ziglar
The Millionaire Next Door by Thomas J. Stanley, PhD
The Power of Positive Thinking by Norman Vincent Peale

Appendix Contents

The items included in this appendix are intended as examples of forms and documents and should be used only upon advice from your attorney.

Appendix A – Partnership Agreement Example

Appendix B – Non-compete Agreement Example

Appendix C – Business Loan Proposal Form Example

Appendix D – Business Plan Example 1

Appendix E – Board of Advisors Agreement Example

Appendix F – Employee Handbook Example

Appendix G – Independent Contractor Agreement Form Example

Appendix H – Business Plan Example 2

Appendix I – ACORD Certificate of Insurance Form Example

A

Partnership Agreement Example

SAMPLE PARTNERSHIP AGREEMENT FORM

YOUR CONSTRUCTION COMPANY, LLC

This Partnership Agreement ("Agreement") is made and entered into as of [Insert Date], by and between Bill Smith and Roy Johnson, hereinafter referred to as the "Members," who hereby form Your Construction Company, LLC (the "Company"), a Georgia Limited Liability Company.

1. This Agreement governs the partnership formed under the laws of the State of Georgia, effective as of the date hereof.
2. The name of the LLC shall be Your Construction Company, LLC.
3. The principal place of business of the Company shall be [Insert Address], or such other place as the Members may from time to time designate.
4. The term of the Company shall commence on the date of this Agreement and shall continue until terminated as provided herein.
5. The purpose of the Company is to engage in any lawful act or activity for which limited liability companies may be organized under the laws of Georgia.
6. The Members shall contribute capital to the Company as follows:
   - Bill Smith: $[amount]
   - Roy Johnson: $[amount]. Further contributions shall be made as mutually agreed upon by the Members.
7. Profits and losses shall be distributed to the Members in proportion to their respective percentage of ownership in the Company.

Mastering the Construction Startup: A Business Infrastructure Guide, First Edition. Nick B. Ganaway.
© 2025 John Wiley & Sons, Inc. Published 2025 by John Wiley & Sons, Inc.

168 | *A Partnership Agreement Example*

8. The business and affairs of the Company shall be managed by the Members. Each Member shall have equal rights in the management and conduct of the Company's business.

9. Each Member shall have one vote in all matters, requiring a vote of the Members. All decisions shall be made by majority vote.

10. No Member may transfer their interest in the Company without first offering such interest to the other Member. The offering Member must provide written notice of their intent to sell their interest, including the terms of the sale, to the remaining Member. The remaining Member shall have a right of first refusal and may choose to purchase the interest under the terms specified in the notice within thirty (30) days of receiving the notice. If the remaining Member declines to purchase the interest, the offering Member may sell their interest to a third party; however, the sale to the third party is subject to the remaining Member's right to approve the potential buyer, which approval shall not be unreasonably withheld.

11. The Company may be dissolved upon the written consent of both Members. In the event that the Members are unable to agree on the dissolution of the Company, the matter shall be resolved through mediation. If mediation fails to resolve the disagreement, the dispute shall be submitted to binding arbitration in accordance with the rules of the American Arbitration Association. The decision of the arbitrator(s) on the matter of dissolution shall be final and binding on both Members.

12. In the event of the death or incapacity of a Member, the remaining Member may elect to continue the business of the Company. Should both Members die or become incapacitated simultaneously or approximately so, the control of the Company shall pass to a designated successor or executor as specified in each Member's will or other estate planning document. In the absence of such specification, the Company shall be managed by a temporary appointed executor or trustee until a permanent solution is established by the court. The Company may also be dissolved according to the laws of the State of Georgia if no viable succession plan is in place.

13. Any disputes under this Agreement shall be resolved by mediation, and if not resolved through mediation, then by binding arbitration.

14. This Agreement shall be governed by and construed in accordance with the laws of the State of Georgia.

15. This Agreement may only be amended by a written agreement signed by all Members.

A *Partnership Agreement Example* | **169**

IN WITNESS WHEREOF, the Members have executed this Partnership Agreement as of the day and year first above written.

[Signature Page Follows]

_____

Bill Smith

_____

Roy Johnson

_____

B

Non-compete Agreement Example

This example non-compete agreement should not be used without review and approval by a lawyer

This Non-compete Agreement ("Agreement") is made effective as of February 1, 2027, by and between Your Construction Company, LLC, a Georgia Limited Liability Company with its principal office located at 123 Main Street, Anytown, Georgia ("Company"), and John Smith ("Employee").

WHEREAS, Employee is employed by Company in a managerial or executive capacity, and in consideration of the Employee's employment with the Company, the parties agree to enter into this Agreement;

WHEREAS, Company wishes to protect its legitimate business interests including its confidential information, trade secrets, and business relationships;

NOW, THEREFORE, in consideration of the mutual promises, covenants, and agreements contained herein, the parties agree as follows:

Non-competition: a. During the term of Employee's employment with the Company and for a period of six (6) months immediately following the termination of employment, whether voluntary or involuntary, Employee agrees not to engage in any employment, consultation, or other activity involving competitive businesses that operate within seventy-five (75) miles of the Company's headquarters or any geographical area in which the Company conducts its business. b. For the purposes of this Agreement, a "competitor" is defined as any person, group of persons, or entity that engages in or owns or controls any enterprise that significantly competes with the business of the Company.

Mastering the Construction Startup: A Business Infrastructure Guide, First Edition. Nick B. Ganaway.
© 2025 John Wiley & Sons, Inc. Published 2025 by John Wiley & Sons, Inc.

Non-solicitation: a. During the term of this Agreement, the Employee agrees not to solicit or induce any employee of the Company to terminate employment with the Company or to work for any competitor of the Company.

Confidentiality: Employee reaffirms their obligations under the existing Confidentiality Agreement signed between Employee and the Company, which prohibits the unauthorized use or disclosure of the Company's confidential information.

Return of Property: Upon termination of employment, Employee agrees to return all Company documents, materials, tools, equipment, and other properties that were used or developed during the course of employment.

Enforcement: If any provision of this Agreement is found to be unenforceable, the remainder shall be enforced as fully as possible, and the unenforceable provision shall be deemed modified to the limited extent required to permit its enforcement in a manner most closely approximating the intention of the parties.

Governing Law: This Agreement shall be governed by and construed in accordance with the laws of the State of Georgia.

Entire Agreement: This Agreement constitutes the entire agreement between the parties regarding the subject matter hereof and supersedes all prior agreements and understandings, both written and oral, between the parties regarding the subject matter hereof.

IN WITNESS WHEREOF, the parties hereto have executed this Non-compete Agreement as of the date first above written.

[Company Signature]

[Name], [Title]

[Employee Signature]

Frank Smith

C

Business Loan Proposal Form Example

There is no universal loan proposal form; however, the information required by lenders is fairly uniform. The following is a generic example.

May 1, 2031

Loan Officer
Last National Bank
123 Finance Avenue
Atlanta, GA 30303

Subject: Your Construction Company, LLC, Business Loan Proposal for $700,000

Dear Loan Officer,
Please accept our proposal for a business loan for an amount of $700,000.

Purpose of the Loan: The primary purpose of this loan is to support the acquisition of additional construction equipment, business office machines, and furniture, and the hiring of skilled labor required for the projects scheduled for this year. We will also lease and renovate additional office space adjoining the existing.

These expenditures are essential as we look to gradually expand our geographical area of operations. In our five years in business here in Atlanta, we have continuously demonstrated our ability to deliver high-quality construction services.

Mastering the Construction Startup: A Business Infrastructure Guide, First Edition. Nick B. Ganaway.
© 2025 John Wiley & Sons, Inc. Published 2025 by John Wiley & Sons, Inc.

Company Overview: Your Construction Company, LLC, established in 2026, has successfully completed more than 100 light commercial projects. Our commitment to excellence has earned us repeat business and attracted new clients through referrals, ensuring continuing profitably during steady growth.

Financial Information: Attached to this letter are our financial statements including income statements, balance sheets, and cash flow statements for the past three years. These documents reflect our company's solid financial stability, profitability, and strong cash flow management.

Loan Repayment: We propose to repay the loan over a period of seven years, through monthly installments derived from our operational cash flows. We have projected our cash flow forecasts, factoring in the additional income from anticipated projects. This ensures our ability to meet repayment terms while maintaining healthy business operations.

Collateral: To secure this loan, we propose that the loan will be guaranteed by Your Construction Company, LLC. We believe you will agree upon review of our financial statements that this provides substantial security for the loan in question.

Conclusion: We are confident in our business plan and our ability to manage this proposed expansion with the same diligence and responsibility that we have demonstrated throughout our company's history. Your Construction Company, LLC, looks forward to the possibility of partnering with Last National Bank to realize the potential growth and opportunities ahead.

Thank you for considering our loan application. We hope to continue to build on our excellent financial relationship with Last National Bank.

Sincerely,
Thomas Williams, President
Your Construction Company, LLC
100 Square Circle
Any City, USA

D

Business Plan Example 1

Business Plan for Your Construction Company, LLC

Date: March 1, 2027

Executive Summary

Company Name: Your Construction Company, LLC
Business Structure: Limited Liability Company (LLC)
Location: Atlanta, GA
Founded: February 1, 2026
Specialization: Construction of buildings with project costs ranging from $1,000,000 to $10,000,000 for general commercial use.

Management Team:

- **Thomas Williams, CEO:** 15 years of construction industry experience.
- **Bill Smith, Senior Project Manager:** 10 years of project management experience.
- **Mary Jones, CPA and Financial Manager:** 12 years of financial management experience.
- **Charles Johnson, Estimator:** 4 years of estimating experience.

Mission Statement: Your Construction Company, LLC is dedicated to delivering high-quality commercial construction services, ensuring client satisfaction through our commitment to excellence, integrity, and innovation.

Mastering the Construction Startup: A Business Infrastructure Guide, First Edition. Nick B. Ganaway.
© 2025 John Wiley & Sons, Inc. Published 2025 by John Wiley & Sons, Inc.

Business Description

Your Construction Company, LLC specializes in the construction of commercial buildings within the Atlanta metropolitan area. Our focus is on projects valued between $1,000,000 and $10,000,000, catering to businesses seeking reliable and professional construction services. We provide a range of services including general contracting, design-build services, project management, and construction consulting.

Founder's Vision

The founder's initial quest in developing this business plan was to determine whether the targeted market offered sufficient business potential and resources to allow Your Construction Company, LLC to thrive now and well into the future. By conducting thorough market research and analysis, it became evident that the commercial construction industry in Atlanta presents significant growth opportunities. This plan also serves a secondary purpose: to demonstrate to the company's lenders and other interested parties the solid foundation upon which Your Construction Company, LLC is built. By showcasing our strategic planning, experienced management team, and detailed financial projections, we aim to instill confidence in our stakeholders and secure the necessary support for our business endeavors.

Competitive Advantage

Your Construction Company, LLC stands out in the competitive commercial construction market due to our experienced management team with a proven track record, strong relationships with local subcontractors and suppliers, and an unwavering commitment to quality and on-time delivery. Our emphasis on safety and regulatory compliance ensures that all projects meet the highest industry standards, providing peace of mind to our clients.

Target Market

Our target market includes small- to medium-sized businesses, property developers, and government agencies in the Atlanta metropolitan area. We aim to serve clients who require high-quality commercial construction services for office spaces, retail outlets, and industrial facilities. By focusing on this specific market segment, we can tailor our services to meet the unique needs of these clients, ensuring their satisfaction and repeat business.

Marketing Plan

To reach our target market, Your Construction Company, LLC will employ a multifaceted marketing strategy. We will engage in networking with industry professionals and attend trade shows to build relationships and showcase our expertise. A strong online presence through a professional website and active social media profiles will help us reach a wider audience and generate leads. Additionally, we will leverage relationships with local real estate agents and architects to gain referrals and increase our visibility in the industry. Offering competitive pricing and flexible payment terms will further attract clients and set us apart from our competitors.

Startup Costs

The total startup costs for Your Construction Company, LLC amount to $1,000,000. This includes $500,000 for the purchase of essential equipment and tools necessary for our construction projects. Office setup and supplies will require an additional $50,000. The remaining $450,000 will be allocated as initial working capital to cover operational expenses and ensure smooth business operations during the initial phase. To fund these startup costs, we seek a loan of $1,000,000, which will enable us to purchase equipment, set up office space, and hire key personnel.

Market Analysis

The commercial construction industry in Atlanta is experiencing steady growth, driven by increasing demand for office spaces, retail outlets, and industrial facilities. Our market analysis indicates that there is a significant opportunity for Your Construction Company, LLC to capture a share of this growing market by leveraging our competitive advantages and targeted marketing efforts.

Organizational Structure

The key personnel in Your Construction Company, LLC include Thomas Williams, CEO; Bill Smith, Senior Project Manager; Mary Jones, CPA and

178 | D Business Plan Example 1

Financial Manager; and Charles Johnson, Estimator. Our team also includes administrative assistants, construction supervisors, skilled laborers, and tradespeople. Each team member brings valuable expertise and experience to the company, ensuring that we can deliver high-quality construction services to our clients.

Financial Plan

Revenue Projections:
- Year 1: $2,500,000
- Year 2: $4,000,000
- Year 3: $6,000,000

Profit Margin:
- Year 1: 10%
- Year 2: 12%
- Year 3: 15%

Break-even Analysis:
- Break-even point: $2,000,000 in annual revenue

Financial Statements:

- **Income Statement:** Detailed projections of revenue, expenses, and profit for the first three years.
- **Cash Flow Statement:** Projected cash inflows and outflows for the first three years.
- **Balance Sheet:** Overview of assets, liabilities, and equity.

Risk Analysis

Potential risks for Your Construction Company, LLC include market fluctuations and economic downturns, delays in project completion, and rising material and labor costs. To mitigate these risks, we will diversify our client base to reduce dependence on a single market segment, establish strong contractual agreements with clients and subcontractors, and maintain a contingency fund to address unexpected expenses.

Conclusion

Your Construction Company, LLC is well positioned to capitalize on the growing demand for commercial construction services in Atlanta. With a skilled management team, a clear market strategy, and a strong financial plan, we are confident in our ability to achieve our business goals and deliver value to our clients and stakeholders. We seek your support and partnership in this endeavor.

This business plan example provides a comprehensive overview of Your Construction Company, LLC, serving as a template for startup entrepreneurs in the commercial construction industry.

E

Board of Advisors Agreement Example

Purpose

The purpose of this agreement is to outline the terms and conditions under which [Your Company Name], located at [Company Address], will engage the services of [Advisor Name] as an advisor to the Board of Advisors.

Parties Involved

Company: [Your Company Name], [Company Address]
Advisor: [Advisor Name], [Advisor Contact Information]

Term of Service

Duration: The advisor's term will commence on [Start Date] and end on [End Date].
Renewal: The term may be renewed upon mutual agreement of both parties.

Roles and Responsibilities

Advisory Role: The advisor will provide expertise and strategic advice in the areas of [Specify Areas].
Meetings: The advisor will attend quarterly meetings, either in person or virtually, as scheduled by the company.

Mastering the Construction Startup: A Business Infrastructure Guide, First Edition. Nick B. Ganaway.
© 2025 John Wiley & Sons, Inc. Published 2025 by John Wiley & Sons, Inc.

Compensation

Payment: The advisor will receive a stipend of [$X] per meeting attended. Reimbursement: The advisor will be reimbursed for reasonable travel expenses incurred in the course of their duties.

Confidentiality

Confidential Information: All information disclosed to the advisor is considered confidential.
Non-disclosure: The advisor agrees not to disclose any confidential information during or after their term.

Conflict of Interest

Disclosure: The advisor must disclose any potential conflicts of interest. Resolution: The company and advisor will mutually resolve any disclosed conflicts.

Indemnification

The company agrees to indemnify and hold the advisor harmless from any legal claims related to their advisory role.

Termination

Termination by Company: The company may terminate this agreement with 30 days' notice.
Termination by Advisor: The advisor may resign from their role with 30 days' notice.

Miscellaneous

Governing Law: This agreement will be governed by the laws of the State of [State].

Entire Agreement: This document constitutes the entire agreement between the parties.

Amendments: Any amendments to this agreement must be made in writing and signed by both parties.

Signatures

For the Company:

Name: [Company Representative Name]

Title: [Title]

Signature: _____

Date: _____

For the Advisor:

Name: [Advisor Name]

Signature: _____

Date: _____

F

Employee Handbook Example

Creating an employee handbook tailored for a small construction-business enterprise involves detailing policies and procedures that ensure smooth operations, compliance with legal requirements, and a positive workplace culture. Below is a sample employee handbook that should fit the needs of "Your Construction Company, LLC."

Your Construction Company, LLC Employee Handbook

Table of Contents

1. **Welcome Message**
2. **Company Overview**
3. **Employment Policies**
   - Equal Employment Opportunity
   - At-Will Employment
   - Employment Classification
4. **Workplace Conduct**
   - Code of Conduct
   - Antiharassment and Nondiscrimination Policy
5. **Compensation and Benefits**
   - Payroll Procedures
   - Overtime Pay
   - Benefits Overview

Mastering the Construction Startup: A Business Infrastructure Guide, First Edition. Edited by Nick B Ganaway.
© 2025 John Wiley & Sons, Inc. Published 2025 by John Wiley & Sons, Inc.

6. Work Hours and Attendance
- Work Schedule
- Attendance and Punctuality
- Timekeeping

7. Leave Policies
- Paid Time Off (PTO)
- Holidays
- Family and Medical Leave (FMLA)

8. Health and Safety
- Safety Program
- Reporting Injuries
- Emergency Procedures

9. Employee Performance
- Performance Reviews
- Disciplinary Procedures

10. Separation of Employment
- Voluntary Resignation
- Involuntary Termination

11. Additional Resources
- Contact Information
- Policy Acknowledgment Form

---

1. Welcome Message

Welcome to Your Construction Company, LLC! We are excited to have you as part of our team. This handbook is designed to provide you with an understanding of our company policies, procedures, and benefits. If you have any questions, please do not hesitate to reach out to your supervisor or the HR department.

2. Company Overview

Your Construction Company, LLC specializes in constructing buildings for general commercial use with project costs ranging from $1,000,000 to $10,000,000. We are dedicated to delivering high-quality construction services and maintaining a safe and inclusive workplace.

3. Employment Policies

Equal Employment Opportunity: We are committed to providing equal employment opportunities to all employees and applicants without regard to race, color, religion, gender, sexual orientation, national origin, age, disability, or any other protected status.

At-Will Employment: Employment with Your Construction Company, LLC is at-will. This means that either you or the company may terminate the employment relationship at any time, with or without cause or notice.

Employment Classification:

- Full-time employees
- Part-time employees
- Temporary employees
- Independent contractors

4. Workplace Conduct

Code of Conduct: Employees are expected to conduct themselves professionally and ethically at all times. This includes honesty, integrity, and respect for others.

Antiharassment and Nondiscrimination Policy: We are committed to maintaining a work environment free from harassment and discrimination. Any form of harassment or discrimination based on race, color, religion, gender, sexual orientation, national origin, age, disability, or any other protected status will not be tolerated.

5. Compensation and Benefits

Payroll Procedures: Employees are paid bi-weekly. Paychecks will be directly deposited into your designated bank account.

Overtime Pay: Nonexempt employees will be paid overtime at a rate of 1.5 times their regular pay for hours worked over 40 in a workweek.

Benefits Overview: We offer a comprehensive benefits package, including health insurance, retirement plans, and paid time off.

6. Work Hours and Attendance

Work Schedule: Standard work hours are from 8:00 AM to 5:00 PM, Monday through Friday. Specific schedules may vary based on project needs.

Attendance and Punctuality: Regular attendance and punctuality are essential. Employees are expected to be at their workstations and ready to work at their scheduled start time.

Timekeeping: Employees are responsible for accurately recording their work hours. Falsifying time records is a serious offense.

7. Leave Policies

Paid Time Off (PTO): Employees accrue PTO based on their length of service and employment classification. PTO can be used for vacation, sick leave, or personal time.

Holidays: We observe the following holidays: New Year's Day, Memorial Day, Independence Day, Labor Day, Thanksgiving Day, and Christmas Day.

Family and Medical Leave (FMLA): Eligible employees may take up to 12 weeks of unpaid, job-protected leave for certain family and medical reasons.

8. Health and Safety

Safety Program: We are committed to providing a safe workplace. Employees are required to follow all safety rules and regulations and to report any unsafe conditions or practices.

Reporting Injuries: All work-related injuries must be reported immediately to your supervisor. An incident report must be completed.

Emergency Procedures: In case of an emergency, follow the emergency procedures posted in your work area.

9. Employee Performance

Performance Reviews: Performance reviews are conducted annually. These reviews provide an opportunity to discuss your job performance, goals, and career development.

Disciplinary Procedures: Employees who violate company policies may be subject to disciplinary action, up to and including termination.

10. Separation of Employment

Voluntary Resignation: Employees who wish to resign are asked to provide at least two weeks' notice in writing.

Involuntary Termination: The company may terminate employment for any lawful reason, including poor performance, misconduct, or organizational needs.

11. Additional Resources

Contact Information: For any questions or concerns, please contact your supervisor or the HR department.

Policy Acknowledgment Form: Employees are required to sign the Policy Acknowledgment Form to confirm they have read and understood the employee handbook.

By signing below, I acknowledge that I have received and read the Your Construction Company, LLC Employee Handbook.

Employee Signature: _____ Date: _____

Supervisor Signature: _____ Date: _____

---

190 | *F Employee Handbook Example*

This sample handbook provides a solid foundation for the policies and procedures of a small construction business enterprise. Adjustments can be made based on specific needs and legal requirements relevant to your company and location.

G

Independent Contractor Agreement Form Example

This agreement form must be completed in accordance with your specific conditions and the laws of your state. Get legal advice before using.

This Independent Contractor Agreement ("Agreement") is made effective as of July 01, 2026, by and between Your Construction Company, LLC ("Company"), of 100 Round Circle, Any City, Georgia, and Cleveland Thompson ("Independent Contractor"), of Loganville, Georgia.

Description of Services. Beginning on July 01, 2026, the Independent Contractor will provide the following services (collectively, "Services"):

Provide supervision, tools, materials, labor, and equipment necessary to clean construction sites and haul off debris on specified projects.

Furthermore, the Independent Contractor has the right of control over how the Independent Contractor will perform the Services. The Company does not have this right of control over how the Independent Contractor will perform the Services.

Payment for Services. The Company will pay compensation to the Independent Contractor for the Services. Payments will be made as follows:

The amount will be determined on a per-job basis in advance and paid upon completion of each assigned job.

No other fees and/or expenses will be paid to the Contractor unless such fees and/or expenses have been approved in advance by the appropriate executive on behalf of the Company in writing. The Independent Contractor shall be solely responsible for any and all taxes, Social Security contributions or

---

Mastering the Construction Startup: A Business Infrastructure Guide, First Edition. Nick B. Ganaway.
© 2025 John Wiley & Sons, Inc. Published 2025 by John Wiley & Sons, Inc.

payments, disability insurance, unemployment taxes, and other payroll-type taxes applicable to such compensation. The Independent Contractor has the right of control over the method of payment for Services.

Term/Termination. Termination of this agreement will occur as follows:

This agreement will terminate upon completion of the assigned project unless terminated by the Company for cause.

Furthermore, the Independent Contractor has the ability to terminate this Agreement "at will."

A regular, ongoing relationship of indefinite term is not contemplated. The Company has no right to assign Services to the Independent Contractor other than as specifically contemplated by this Agreement. However, the parties may mutually agree that the Independent Contractor shall perform other services for the Company, pursuant to the terms of this Agreement.

Relationship of Parties. It is understood by the parties that the Independent Contractor is an independent contractor with respect to the Company and not an employee of the Company. The Company will not provide fringe benefits, including health insurance benefits, paid vacation, or any other employee benefit, for the benefit of the Independent Contractor.

It is contemplated that the relationship between the Independent Contractor and the Company shall be a nonexclusive one. The Independent Contractor also performs services for other organizations and/or individuals. The Company has no right to further inquire into the Independent Contractor's other activities.

Company's Control. The Company has no right or power to control or otherwise interfere with the Independent Contractor's mode of effecting performance under this Agreement. The Company's only concern is the result of the Independent Contractor's work, and not the means of accomplishing it. Except in extraordinary circumstances and when necessary, the Independent Contractor shall perform the Services without direct supervision by the Company.

Professional Capacity. The Independent Contractor is a professional who uses their own professional and business methods to perform Services.

The Independent Contractor has not and will not receive training from the Company regarding how to perform the Services.

Personal Services Not Required. The Independent Contractor is not required to render the Services personally and may employ others to perform the Services on behalf of the Company without the Company's knowledge or consent. If the Independent Contractor has assistants, it is the Independent Contractor's responsibility to hire them and to provide materials for them.

No Location on the Premises. The Independent Contractor has no desk or other equipment either located at or furnished by the Company. Except to the extent that the Independent Contractor works in a territory as defined by the Company, their Services are not integrated into the mainstream of the Company's business.

No Set Work Hours. The Independent Contractor has no set hours of work. There is no requirement that the Independent Contractor work full time or otherwise account for work hours.

Expenses Paid by Independent Contractor. The Independent Contractor's business and travel expenses are to be paid by the Independent Contractor and not by the Company.

Confidentiality. The Independent Contractor may have had access to proprietary, private, and/or otherwise confidential information ("Confidential Information") of the Company. Confidential Information shall mean all nonpublic information that constitutes, relates, or refers to the operation of the business of the Company, including without limitation, all financial, investment, operational, personnel, sales, marketing, managerial, and statistical information of the Company, and any and all trade secrets, customer lists, or pricing information of the Company. The nature of the information and the manner of disclosure are such that a reasonable person would understand it to be confidential. The Independent Contractor will not at any time or in any manner, either directly or indirectly, use for the personal benefit of the Independent Contractor, or divulge, disclose, or communicate in any manner any Confidential Information. The Independent Contractor will protect such information and treat the

Confidential Information as strictly confidential. This provision shall continue to be effective after the termination of this Agreement. Upon termination of this Agreement, the Independent Contractor will return to the Company all Confidential Information, whether physical or electronic, and other items that were used, created, or controlled by the Independent Contractor during the term of this Agreement.

This Agreement is in compliance with the Defend Trade Secrets Act and provides civil or criminal immunity to any individual for the disclosure of trade secrets: (i) made in confidence to a federal, state, or local government official, or to an attorney when the disclosure is to report suspected violations of the law; or (ii) in a complaint or other document filed in a lawsuit if made under seal.

Injuries. The Independent Contractor acknowledges the Independent Contractor's obligation to obtain appropriate insurance coverage for the benefit of the Independent Contractor (and the Independent Contractor's employees, if any). The Independent Contractor waives any rights to recovery from the Company for any injuries that the Independent Contractor (and/or the Independent Contractor's employees) may sustain while performing the Services under this Agreement and that are a result of the negligence of the Independent Contractor or the Independent Contractor's employees. The Independent Contractor will provide the Company with a certificate naming the Company as an additional insured party.

Indemnification. The Independent Contractor agrees to indemnify and hold harmless the Company from all claims, losses, expenses, fees including attorney fees, costs, and judgments that may be asserted against the Company that result from the acts or omissions of the Independent Contractor, the Independent Contractor's employees, if any, and the Independent Contractor's agents.

No Right to Act as Agent. An "employer–employee" or "principal–agent" relationship is not created merely because (1) the Company has or retains the right to supervise or inspect the work as it progresses in order to ensure compliance with the terms of the Agreement; or (2) the Company has or retains the right to stop work done improperly. The Independent Contractor has no right to act as an agent for the Company and has an obligation to notify any involved parties that it is not an agent of the Company.

Independent Contractor Agreement Form Example | **195**

Entire Agreement. This Agreement constitutes the entire agreement between the parties. All terms and conditions contained in any other writings previously executed by the parties regarding the matters contemplated herein shall be deemed to be merged herein and superseded hereby. No modification of this Agreement shall be deemed effective unless in writing and signed by the parties hereto.

Waiver of Breach. The waiver by the Company of a breach of any provision of this Agreement by the Independent Contractor shall not operate or be construed as a waiver of any subsequent breach by the Independent Contractor.

Severability. If any provision of this Agreement shall be held to be invalid or unenforceable for any reason, the remaining provisions shall continue to be valid and enforceable. If a court finds that any provision of this Agreement is invalid or unenforceable, but that by limiting such provision it would become valid and enforceable, then such provision shall be deemed to be written, construed, and enforced as so limited.

Applicable Law. This Agreement shall be governed by the laws of Georgia.

Signatories. This Agreement shall be signed by Tom Williams, Owner on behalf of Your Construction Company, LLC, and by Cleveland Black. This Agreement is effective as of the date first above written.

The Company:

Your Construction Company, LLC

By: ___ Date_____

Tom Williams, Owner

The Independent Contractor:

By: ___ Date_____

Cleveland Thompson, Independent Contractor

H

Business Plan Example 2

1. Company Overview

Name and Location: Your Construction Company, LLC, Atlanta, GA
Legal Structure: Limited Liability Company (LLC)
Founding Date: February 1, 2026
Mission Statement: To build enduring value for clients and communities through innovative, sustainable commercial construction that exceeds expectations.

2. Core Values

Integrity: Committing to the highest standards of professionalism and ethical conduct.
Quality: Delivering excellence in every project through attention to detail and best practices.
Safety: Prioritizing the health and safety of our workers and stakeholders at all sites.
Teamwork: Cultivating collaboration and mutual respect among our team and partners.
Sustainability: Embracing environmentally responsible construction practices.
Services Offered: Specializing in the construction of commercial buildings, with project costs ranging from $1,000,000 to $10,000,000.

Mastering the Construction Startup: A Business Infrastructure Guide, First Edition. Nick B. Ganaway.
© 2025 John Wiley & Sons, Inc. Published 2025 by John Wiley & Sons, Inc.

3. Management Team

Thomas Williams, Founder and CEO: Provides vision and overall strategic management.

Bill Smith, Sr. Project Manager: Along with three other project managers, oversees the planning and management of construction projects.

Mary Jones, CPA, Financial Manager: Manages financial operations and planning.

Charles Johnson, Estimator: Responsible for project cost estimations.

4. Market Analysis

Industry Overview: The commercial construction industry in Atlanta has seen steady growth due to economic expansion and increased demand for commercial space. The industry is competitive but offers significant opportunities for firms that prioritize quality and efficiency.

Target Market: Our primary market includes real estate developers, large enterprises, and government entities looking to develop new commercial properties or expand existing ones.

Competitive Analysis: The market consists of several large firms that compete on scale but often lack the personalized service smaller firms like Your Construction Company, LLC, can provide. Our focus on customer service and sustainability differentiates us from competitors.

5. Marketing Strategy

Branding: We will position ourselves as a leader in sustainable commercial construction, emphasizing our commitment to quality and environmental stewardship.

Outreach: Marketing efforts will include networking within local and regional real estate and business communities, online marketing through a professional website and social media, and participation in industry conferences.

Customer Retention: We will maintain high customer satisfaction through continuous engagement during and after the project completion to ensure all client needs are met.

6. Operations Plan

Facilities: Our main office is located in Atlanta, GA, with easy access to most project sites and ample space for our administrative staff and planning teams.
Processes: From initial bids to project completion, our operations are characterized by careful attention to quality standards and efficiency. Regular training sessions ensure our team remains adept in the latest construction techniques and safety protocols.

7. Financial Plan

Startup Costs: Initial funding was directed toward securing a lease for our office space, purchasing essential equipment, and initial marketing activities to establish our brand presence.
Projected Financials: Detailed financial projections (not included here) will outline expected revenues, cost of goods sold, operating expenses, and net profit margins over the next five years.
Funding Requirements: We are seeking through Last National Bank in Atlanta an additional $700,000 in business loans to finance upcoming projects and equipment upgrades.

8. Goals and Objectives

Short-term Goals: Secure at least three suitable project contracts in 2026, continuously monitor and improve our operational processes, and expand our project management team.
Long-term Goals: Become a leading commercial construction firm among our category of projects in Atlanta focused on sustainable building practices within five years.

9. Risk Management

Identified Risks: Economic downturns affecting project funding, delays in supply chains, and demand for new construction products.
Mitigation Strategies: Maintaining strong financial reserves and availability by way of good banking relationships, diversifying supplier relationships, and adhering to strict safety protocols.

ACORD Certificate of Insurance Form Example

ACORD®	**CERTIFICATE OF LIABILITY INSURANCE**	**DATE (MM/DD/YYYY)** 05/01/2026

THIS CERTIFICATE IS ISSUED AS A MATTER OF INFORMATION ONLY AND CONFERS NO RIGHTS UPON THE CERTIFICATE HOLDER. THIS CERTIFICATE DOES NOT AFFIRMATIVELY OR NEGATIVELY AMEND, EXTEND OR ALTER THE COVERAGE AFFORDED BY THE POLICIES BELOW. THIS CERTIFICATE OF INSURANCE DOES NOT CONSTITUTE A CONTRACT BETWEEN THE ISSUING INSURER(S), AUTHORIZED REPRESENTATIVE OR PRODUCER, AND THE CERTIFICATE HOLDER.

IMPORTANT: If the certificate holder is an ADDITIONAL INSURED, the policy(ies) must have ADDITIONAL INSURED provisions or be endorsed. If SUBROGATION IS WAIVED, subject to the terms and conditions of the policy, certain policies may require an endorsement. A statement on this certificate does not confer rights to the certificate holder in lieu of such endorsement(s).

PRODUCER		CONTACT NAME:	Whatever Agent		
Whatever Insurance Broker		PHONE (A/C, No, Ext): (555) 203-2030		FAX (A/C, No):	
Street number and name		E-MAIL ADDRESS: whateveragent@email.com			
City, State and Zip		INSURER(S) AFFORDING COVERAGE			NAIC #
(xxx) xxx-xxxx		INSURER A: Best Insurance Company #1			29897
INSURED		INSURER B: Other Insurance Company #2			34469
Contractor Company Name		INSURER C: Third Insurance Company			23975
Contractor street address		INSURER D:			
Contractor City, State and Zip		INSURER E:			
		INSURER F:			

COVERAGES   CERTIFICATE NUMBER: 1234567   REVISION NUMBER:

THIS IS TO CERTIFY THAT THE POLICIES OF INSURANCE LISTED BELOW HAVE BEEN ISSUED TO THE INSURED NAMED ABOVE FOR THE POLICY PERIOD INDICATED. NOTWITHSTANDING ANY REQUIREMENT, TERM OR CONDITION OF ANY CONTRACT OR OTHER DOCUMENT WITH RESPECT TO WHICH THIS CERTIFICATE MAY BE ISSUED OR MAY PERTAIN, THE INSURANCE AFFORDED BY THE POLICIES DESCRIBED HEREIN IS SUBJECT TO ALL THE TERMS, EXCLUSIONS AND CONDITIONS OF SUCH POLICIES. LIMITS SHOWN MAY HAVE BEEN REDUCED BY PAID CLAIMS.

INSR LTR	TYPE OF INSURANCE	ADDL INSD	SUBR WVD	POLICY NUMBER	POLICY EFF (MM/DD/YYYY)	POLICY EXP (MM/DD/YYYY)	LIMITS	
A	☒ COMMERCIAL GENERAL LIABILITY	N	N	AB23456	5/01/2026	05/01/2027	EACH OCCURRENCE	$ 750,000
	☐ CLAIMS-MADE ☒ OCCUR						DAMAGE TO RENTED PREMISES (Ea occurrence)	$ 500,000
							MED EXP (Any one person)	$ EXCLUDED
							PERSONAL & ADV INJURY	$ 750,000
	GEN'L AGGREGATE LIMIT APPLIES PER:						GENERAL AGGREGATE	$ 1,750,000
	☐ POLICY ☐ PROJECT ☐ LOC						PRODUCTS - COMP/OP AGG	$ 1,750,000
	☐ OTHER:							$
A	AUTOMOBILE LIABILITY	N	N	KYB23456	05/01/2026	05/01/2027	COMBINED SINGLE LIMIT (Ea accident)	$ 2,000,000
	☒ ANY AUTO						BODILY INJURY (Per person)	$ XXXXXXX
	☐ OWNED AUTOS ONLY ☐ SCHEDULED AUTOS						BODILY INJURY (Per accident)	$ XXXXXXX
	☒ HIRED AUTOS ONLY ☒ NON-OWNED AUTOS ONLY						PROPERTY DAMAGE (Per accident)	$ XXXXXXX
								$ XXXXXXX
B	☒ UMBRELLA LIAB ☒ OCCUR	N	N	04523456	05/01/2026	05/01/2027	EACH OCCURRENCE	$ 25,000,000
	☐ EXCESS LIAB ☐ CLAIMS-MADE						AGGREGATE	$ 25,000,000
	☐ DED ☐ RETENTION $							$ XXXXXXX
A	WORKERS COMPENSATION AND EMPLOYERS' LIABILITY Y/N	N/A	N	MPQ23456	05/01/2026	05/01/2027	☒ PER STATUTE ☐ OTHER	
	ANY PROPRIETOR/PARTNER/EXECUTIVE OFFICER/MEMBER EXCLUDED? ☐ n (Mandatory in NH)						E.L. EACH ACCIDENT	$ 1,000,000
							E.L. DISEASE - EA EMPLOYEE	$ 1,000,000
	If yes, describe under DESCRIPTION OF OPERATIONS below						E.L. DISEASE - POLICY LIMIT	$ 1,000,000
C	Property	N	N	PPR23456	05/01/2026	05/01/2027	$10,000,000 Per Occurrence *See attached	

DESCRIPTION OF OPERATIONS/LOCATIONS/VEHICLES (ACORD 101, Additional Remarks Schedule, may be attached if more space is required)

RE: (Name property type here) Unit #1234 @ 555 Main St., Everytown, USA 12345.
Certificate Holder is included as Additional Insured as respects location listed for General Liability and Umbrella Coverage when required by written agreement with the insured prior to a loss.

CERTIFICATE HOLDER	CANCELLATION
Company Name Company Street number and name Company city, state and zip code	SHOULD ANY OF THE ABOVE DESCRIBED POLICIES BE CANCELLED BEFORE THE EXPIRATION DATE THEREOF, NOTICE WILL BE DELIVERED IN ACCORDANCE WITH THE POLICY PROVISIONS.
	AUTHORIZED REPRESENTATIVE Agent signature

© 1988–2015 ACORD CORPORATION. All rights reserved

ACORD 25 (2016/03)   The ACORD name and logo are registered marks of ACORD

ACORD certificate of liability insurance. Source: ACORD Corporation.

Mastering the Construction Startup: A Business Infrastructure Guide, First Edition. Nick B. Ganaway.
© 2025 John Wiley & Sons, Inc. Published 2025 by John Wiley & Sons, Inc.

Index

Note: Page numbers in **bold** refers to tables.

a

accounting   36, 78, 91–96, 138
ACORD certificate of insurance form
    example   122, 201
adaptability   13, 105
American Institute of Architects
    (AIA)   72, 76
arbitration   65–66
architecture, engineering, and
    construction industry (A/E/C
    industry)   73
assets   20–21, 23, 95, 110, 117, 156
  business   20, 121
  company's   21, 95
  personal   20, 111, 144
Associated Builders and Contractors
    (ABC)   44, 72, 76
Associated General Contractors
    (AGC)   44, 72, 76
automotive stores   133–135

b

balance sheet   95, 110, 178
banking   109–112

board of advisors   137–139
  agreement example   181–183
budgets, proactively manage   40–41
builders' risk insurance   117
Burger King (BK)   76–77
business and government
    regulations   17
  C Corporation   22–23
  choosing name for company   25
  corporate entity   25
  corporations   19–21
  incorporation   17
  LLCs   19–21
  municipal business licensing and
    permitting   18–19
  partnership   24
  partnership agreement   24–25
  professional licensing   18
  registered agent   18
  registration in other states   17–18
  S Corporation   22
  sole proprietorship   23
business loan proposal form
    example   173–174

Mastering the Construction Startup: A Business Infrastructure Guide, First Edition. Nick B. Ganaway.
© 2025 John Wiley & Sons, Inc. Published 2025 by John Wiley & Sons, Inc.

204 | *Index*

business owner's policy (BOP)   117
business plan   123
  business description   176
  company overview   197
  competitive advantage   176
  for construction company   175
  core values   197
  example   175, 197
  executive summary   175
  financial plan   178, 199
  founder's vision   176
  goals and objectives   199
  management team   175, 198
  market analysis   177, 198
  marketing plan   177
  marketing strategy   198
  mission statement   175
  operations plan   199
  organizational structure   177–178
  presentation for user   123–124
  proposal guidelines   124–125
  purpose and evaluation   123
  risk analysis   178
  risk management   199
  startup costs   177
  target market   176–177
business threats and opportunities   39

C

capital gains tax   155–156
casual dining restaurants   133–134
C corporation (C Corp)   21–23
certificate of insurance
  (COI)   113–114
certified public accountant firm
  (CPA firm)   19–20, 25, 93–94,
  96, 142, 159

chain stores   133, 135–136
  operators   129–130
  owners and franchisors   133–136
change orders   31, 33, 40, 49, 57–59
chief executive officer (CEO)   6
Chrysler   10
civil litigation lawyer   88
claims   19–20, 23–24, 48, 73, 88,
  114–115, 120, 122
  construction industry   63
  delay   67
  high-dollar   61
  IRS, 57
  workers compensation   87, 116
coinsurance   119
commercial insurance companies   113
commercial property insurance   118
communication skills   12
company culture   9–10, 99
ConsensusDocs firm   73
construction bookkeeping software   38
construction business   4–6, 47, 91,
  147, 156
  magazines   44
  owner   161
construction contracts   120
  within chain store market   131
  types of   73
construction disputes   61–66
Construction Dive   161
construction document software   38
construction executive   162
construction industry lawyer   86–87
construction insurance   113
  builders' risk insurance   117
  business owner's policy   117
  coinsurance   119

Index | 205

commercial property insurance 118
cost of workers' comp insurance
116–117
employer responsibilities under
workers comp insurance 116
fidelity insurance 118
general/commercial liability
insurance 114
home-based business insurance 118
making claims 120
performance and payment
bonds 120–121
product liability insurance 117
professional liability insurance 118
workers' compensation
insurance 115–116
construction type 34
contract(s) 20, 48, 65, 71–72, 149
construction 73, 131
nonunion 34
terms and conditions 63
traditional 75
unit price 74
contractor failure
changing geographic area 36
key personnel, change in 36
management maturity, lack of 36–38
managing common causes of 35
new type of construction 36
project size, increase in 35–36
convenience stores 130
corporate entity 25
corporations 19–23, 143
cost of goods sold (COGS) 94
cost of workers' comp insurance
116–117
cost-plus contract 74–75

credit unions 111–112
culture 9–10

d

Daimler-Benz 10
deal making 89–90
design–build contract 75
design defects 61–62
differing conditions 53–55
dispute resolution methods 64–66
due diligence 100, 148–149

e

earnings before interest, taxes, and
amortization (EBITA). see operating
income
economy, awareness of 39
effective change order
procedures 58–59
effective communication 12, 42, 48
effective subcontract
documents 47–59
emotional intelligence (EI) 12
empathy 12
employee handbook 105
additional resources 189–190
company overview 186
compensation and benefits 187–188
employee performance 189
employment policies 187
example 185
health and safety 188
separation of employment 189
welcome message 186
work hours and attendance 188
workplace conduct 187
employee meetings 105

206 | *Index*

employer responsibilities under
  workers comp insurance   116
employment lawyer   87
empowerment   13
Engineering News-Record   162
entrepreneurial characteristics   3
  construction business   4–6
  owner and manager of company
    6–8
  passion   7–8
entrepreneurs   3, 5, 11, 29–30, 98, 147,
    153, 162
estate planning lawyer   89
ethical leadership   13
existential business crisis
  case study of   141
  perfect storm   142–143
  personal effect   144–145
  success   145–146
  tough decision   144

f
family and medical leave
    (FMLA)   186, 188
fidelity insurance   118
financing activities   95
fixed-price contract. *see* lump-sum
    contract
Forbes   162
functional magnetic resonance imaging
    (fMRI)   3

g
general/commercial liability insurance
    (CGL)   114
general contractor   29, 47, 50–52, 55,
    57, 59, 61–62, 67, 70, 76–77, 85, 88,
    94, 129–130, 147, 149

awareness of economy and business
    threats and opportunities   39
construction bookkeeping
    software   38
construction document software   38
construction type   34
cultivate relationships   41–42
demand high quality   41
human resource management   42
insurance   56
license   18
managing common causes of
    contractor failure   35–38
marketing   43–45
negotiation   39–40
owner–contractor
    relationship   30–32
proactively manage budgets   40–41
risk management   32–34
schedule   42
stakeholder communication   42
task dependency software   38
treatment of workers or
    company   55
trust   43
union or nonunion   34–35
general practice or business lawyer
    87
Gen Z'ers   98
Gen Z group   98
gross profit   94
guaranteed maximum price (GMP)   75

h
heating, ventilation and air conditioning
    (HVAC)   55, 132
high quality   41, 43

Index | **207**

hiring process   97
  employee handbook   105
  employee meetings   105
  job benefits   100–102
  jobsite facilities and conveniences
    102–103
  noncompete agreement   106–107
  onboarding new employees   104–105
  right people   98–99
  strong company values, examples
    of **99**
  tips for interviewing prospective
    employees   99–100
home-based business insurance   118
human resource management   42

i

immigration lawyer   89
Inc.   162
incentive contract   74–75
income statement   94–95, 110, 178
incorporation   17, 23
independent contractor   55–56, 106.
    see also general contractor
  agreement form example   191–195
  IRS regulations for   56–57
inspection of subcontractor work   52
integrity   13, 30–31, 93, 120, 175, 187
intellectual property lawyer (IP lawyer)
    89
investing activities   95

j

job benefits   100–102
jobsite facilities and conveniences
    102–103
Journal of Business Venturing   3
Journal of Light Construction   161

l

lawsuits   18–20, 23, 48, 61, 86, 88, 114,
    116, 194
  example   79–81
  outcome of construction   38
lawyers   48, 71, 85–90, 147
  hiring   65
  plaintiffs'   20
  project owner's   72
leadership, elements of   11–13
LegalZoom   18, 25
lenders   35–36, 111
liabilities   24, 95, 117
limited liability company (LLC)   19–21,
    137, 197
loan documentation   110
lump-sum contract   73–75

m

manager of company   6–7
marketing   43–45, 193
  plan   177
  strategy   198
mediation   64–65
mergers & acquisitions lawyer   88
money   4, 19–20, 40, 72, 74, 86,
    153, 162
  magazines   5
  sums of   89
motivation   12, 146
municipal business licensing   18–19

n

nation's economic conditions   98
negligence   62–63, 118, 194
negotiation   39–40, 59, 63–64, 75, 150
net income   95

208 | *Index*

New York Times, The   162
niche contracting   45, 129
  additional niche
    advantages   131–133
  advantages   129–131
  chain store owners and
    franchisors   133–136
noncompete agreement   106–107
  difficult-to-enforce   24
  example   171–172
nonshored trench cave-ins   62
nonunion   34–37
not-to-exceed price (NTE price)   75

o

Occupational Safety and Hazards
  Administration (OSHA)   62
official notice   18, 33, 51
onboarding new employees   104–105
operating activities   95
operating agreement   21, 23
operating expenses   94, 124, 199
operating income   94–95, 149
organizational culture   9
outside board of advisors   137–139
owner–contractor relationship
  30–32
owner of company   6–7
owner's equity   95

p

paid time off (PTO)   56, 101,
  186, 188
partnerships   24, 137
  agreement   21, 24–25, 167–169
  strategic   69
passion   7–8

payment(s)   49, 56, 64, 182
  bonds   120–121
  delayed   33
  disputes   64, 68
  of projects   76–78
  for services   191
  timely   31–32
payroll employee   55–56
performance bonds   120–121
permitting   18–19, 78, 148
personal guarantee   20, 111, 120
personal injury lawyer   89
personal protective equipment
  (PPE)   103
precaution, notice to proceed   51
prequalification of
  subcontractors   49–50
private lenders   112
private loans   112
product liability insurance   117
professional liability insurance   118
professional licensing   18–19
profit and loss statement   94–95
project delay   33, 67
  causes for   67–68
  preparation   69–70
project labor agreements (PLAs)   34
project schedule   32, 38, 42

q

qualified remodeler   161
quick serve restaurants (QSRs)   133

r

real estate investing   147–156
real estate lawyer   88
registered agent   18

Index | 209

registration in other states   17–18
relationships   41–42
resilience   5, 13, 145
revenues   94
risk management   32–34, 199

s

sale-leaseback transaction (SLB
transaction)   148–149
scope of work   49–51, 74
change in   63
S Corporation (S Corp)   19, 22
self-awareness   12
self-regulation   12
shareholders' equity. *see* owner's
equity
Silicon Valley Bank   109
single-tenant net-lease real estate
ownership (STNL real estate
ownership)   147
general characteristics   148–150
marketplace   150
primary factors in pricing   151–153
properties   148
replacing property early   153–154
vs. stock market   154–155
social skills   12, 104
Society for Human Resource
Management (SHRM)   104
sole proprietorships   21, 23, 25,
137, 159
stakeholder communication   42
startup businesses, summary checklist
for   159–160
statement of cash flows   95–96
stock market   147, 154–155
strategic direction   11–12

subcontractor management   47–59
subcontractor proposal   50–51

t

task dependency software   38, 51
tax(es)   91–96, 156
lawyer   88
payroll   55, 57
preparation   92–94
team development   13
1031 tax-deferred exchange   156
topographical site survey   52–53
triple-net lease properties (NNN)
148, 150
trust   13, 43, 48–49, 139
mutual   30, 32
and responsibility   102

u

union   34–36
unit price contract   74
U.S. businesses   19, **19**
US Internal Revenue Service (IRS)
19, 55
audit   93
provision   155
regulations for independent
contractors   56–57

v

values   9–10, 38
of business   24
core   197
examples of strong company   **99**
insured   119
intangible   10
of written agreement   72–73

210 *Index*

variable rate loans 111
vision 11–12, 198

W
Wall Street Journal, The 3, 93, 163
Wired 162
work
  changes in 63
  hours and attendance 188
  inspection of subcontractor 52
  scope of 50–51

stoppages 35
workers compensation (WC)
  benefits to employers 116
  cost of 116–117
  insurance 115–116
  lawyer 87–88
workmanship defects 62
written agreement, value of 72–73

Z
ZenBusiness 18